Progress in Optical Science and Photonics

Volume 10

The purpose of the series Progress in Optical Science and Photonics is to provide a forum to disseminate the latest research findings in various areas of Optics and its applications. The intended audience are physicists, electrical and electronic engineers, applied mathematicians, biomedical engineers, and advanced graduate students.

More information about this series at http://www.springer.com/series/10091

Indrani Bhattacharya · Lakshminarayan Hazra

Azimuthal Walsh Filters

A Tool to Produce 2D and 3D Light Structures

Springer

Indrani Bhattacharya
Department of Physics
and Mathematics, Institute of Photonics
University of Eastern Finland
Joensuu, Finland

Lakshminarayan Hazra
Department of Applied Optics and Photonics
University of Calcutta
Kolkata, West Bengal, India

ISSN 2363-5096 ISSN 2363-510X (electronic)
Progress in Optical Science and Photonics
ISBN 978-981-15-6098-9 ISBN 978-981-15-6099-6 (eBook)
https://doi.org/10.1007/978-981-15-6099-6

This Springer imprint is published by the registered company Springer Nature Singapore Pte Ltd.
The registered company address is: 152 Beach Road, #21-01/04 Gateway East, Singapore 189721, Singapore

Preface

This book explores the possibility of using azimuthal Walsh filters derived from azimuthal Walsh functions as an effective tool for the generation of 2D and 3D light structures by manipulating far-field diffraction characteristics near the focal plane of a rotationally symmetric imaging system when used as pupil filters. Starting with the definition of Walsh functions, this book reports the generation and synthesis of azimuthal Walsh filters. The inherent self-similarity present in the various orders of the azimuthal Walsh filters has been explored, and the filters are classified into self-similar groups and sub-groups. Azimuthal Walsh filters are observed to possess a unique rotational self-similarity exhibited among adjacent orders.

Diffraction characteristics of azimuthal Walsh filters placed at exit pupil plane have been computed analytically for far-field plane and are presented in this book. The transverse intensity distributions along the far-field plane shows asymmetrical patterns due to the phase asymmetries introduced by azimuthal Walsh filters. Transverse intensity distributions on the far-field plane exhibited by each group member have self-similar properties. They correspond to the structural self-similarity of the filters in the group. The effect of defocus or 'Depth of Focus' aberration, in case of shifting of image plane in and around focal plane, has also been presented. The results have been compared and analysed, and the possibility of using them as the basis functions for trapping of contact-free handling of particles of the size of micro- and nano-scale is proposed, where light and matter can possibly interact.

Practical implementation of these filters can be achievable by the availability of high-speed spatial light modulators where the in situ generation of different orders of azimuthal Walsh filters is possible and combinations of lower order filters to achieve higher orders may be explored. The lossless azimuthal Walsh filters are energy efficient and can be used to solve the problem of direct implementation of a finite number of constant as well as varying phase levels. The orthogonality and self-similarity of the azimuthally varying Walsh filters could be harnessed to sculpture 3D light distribution near the focus that could find potential application to different areas of physical sciences, namely of the inverse problem where a phase filter needs to be synthesized in accordance with pre-specified diffraction characteristics.

The authors wish to acknowledge PREIN—Photonics REsearch and INnovation, the Finnish Flagship Program running at the Institute of Photonics and Department of Physics and Mathematics, University of Eastern Finland, Joensuu, Finland, and Department of Applied Optics and Photonics, University of Calcutta, Kolkata, India, for the necessary support to write this monograph.

Joensuu, Finland — Indrani Bhattacharya
Kolkata, India — Lakshminarayan Hazra
April 2020

Contents

About the Authors

Dr. Indrani Bhattacharya is presently associated with the Institute of Photonics, University of Eastern Finland, Joensuu. She was previously associated with Prof. Ayan Banerjee of Light-Matter Interaction Lab, Indian Institute of Science Education and Research, IISER, Kolkata, and Prof. Vasudevan Lakshminarayanan of School of Optometry and Vision Science, University of Waterloo, Canada, as a freelance postdoctoral researcher. She has completed B.Sc. with Hons. in Physics and M.Sc. (Tech.) in Applied Physics with specialization in Optics and Opto-electronics from the University of Calcutta, India. She came back to academia to pursue Ph.D. in Technology from the Department of Applied Optics and Photonics, University of Calcutta after a long period of almost 20 years serving in industry working in fibre optics. Her research areas and interests include diffractive optics, biomimetics, optical tweezers, point spread function engineering, in-vivo and in-vitro biomedical applications, optical fibre sensors. She is the current Chair of Steering Committee Members of International Society of Optomechatronic Technology, ISOT, Member of International Society of Optics and Photonics, (SPIE), European Optical Society (EOS), Optical Society of India (OSI), and serving as the Scholarship Committee Member of SPIE for 2019–2020. She was honoured and enlisted by SPIE in Women in Optics Planner in 2016–2017 for her long-term contribution in Optics and Photonics. She is the Convener and Organizer of several International Conferences in Optics and Photonics and Editor of Springer Proceedings in Physics of Volumes 166, 194, 233, 249 and has published several research articles.

Prof. Lakshminarayan Hazra is a B.Sc. (Hons) Physics Graduate and Post-graduate of Applied Physics with a Doctorate from the University of Calcutta, Kolkata. Prof. Lakshminarayan Hazra has over four decades of academic and industrial experience. He is an Emeritus Professor and Former Head of the Department of Applied Optics and Photonics at the University of Calcutta, Kolkata, India. His areas of professional specialization include lens design/optical system design, image formation and aberration theory, diffractive optics, and optical and photonic instrumentation. He is a Fellow of the Optical Society of America and the

International Society for Optics and Photonics (SPIE). He is the editor in chief of the archival journal, *Journal of Optics*, published by M/s Springer in collaboration with the Optical Society of India. He has published more than 150 journal articles and books.

Chapter 1
Walsh Functions, Walsh Filters and Self-Similarity

1.1 Introduction

The concept of Walsh functions was introduced by J. L. Walsh in 1923 as a closed set of normal orthogonal functions defined over the interval (0, 1) and having the values +1 or −1 within the interval [1]. But until recently these functions have not been used to any significant extent in the treatment of physical problems [2]. The system of Walsh functions has some unique properties [3] that the systems are orthonormal in Hilbert space L_2 (0, 1), and can be derived from the character group of the dyadic group which is a topological group derived from the set of binary presentations of the real numbers. Walsh functions may be defined for a single variable or two variables in rectangular coordinates [4]. Formulations in polar coordinates can be made via orthogonality properties [5–7]. Radial and azimuthal Walsh functions can be derived from the polar Walsh functions [1, 8].

A structure is called self-similar when it can be divided into smaller and smaller pieces when each piece becomes an exact replica of the entire structure preserving the details of all scales, however small. A structure may be called fractal if it possesses self-similarity. The term 'Fractals' was introduced by Benoit B. Mandelbrot in 1975 while developing a theory of "roughness and self-similarity" in nature and later he wrote a book on this [9]. Self-similarity of azimuthal Walsh functions and filters, their divisions into self-similar groups and subgroups and representation of each group by analytical formulae has been discussed in detail in this chapter.

I. Bhattacharya and L. Hazra, *Azimuthal Walsh Filters*, Progress in Optical Science and Photonics 10,
https://doi.org/10.1007/978-981-15-6099-6_1

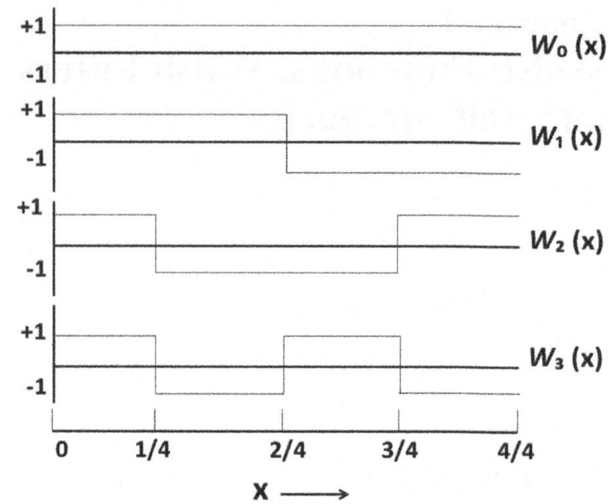

Fig. 1.1 One dimensional Walsh functions, $W_\nu(x)$, $\nu =$ 0, 1,…, 3

1.2 One Dimensional Walsh Functions

A Walsh function of a single variable x may be represented as $W_{k}(x)$ where the subscript k refers to the order of the function. The function takes on values either +1 or -1 over a specified domain and the set of Walsh function is orthogonal and complete in the interval. The order k of the function represents the number of zero crossings or phase transitions within the interval. The orthogonality condition for the set of Walsh functions defined for the interval (0,1) implies the following relation,

$$\int_0^1 W_\nu(x)\, W_{\text{ƀ}}(x)dx = \delta_{\nu \text{ƀ}} \tag{1.1}$$

where $\delta_{\nu \text{ƀ}}$ is the Krönecker delta defined as

$$\delta_{\nu \text{ƀ}} = \begin{cases} 0, & \nu \neq \text{ƀ} \\ 1, & \nu = \text{ƀ} \end{cases} \tag{1.2}$$

The first four one-dimensional Walsh Functions in rectangular co-ordinates are shown in Fig. 1.1.

1.3 Two Dimensional Walsh Functions

1.3.1 Rectangular Walsh Functions

In two dimensions, the orthogonality condition takes the form

$$\int_0^1 \int_0^1 W_{ƙ}(x)\, W_{\nu}(y) W_{€}(x) W_{£}(y) dxdy = \delta_{ƙ€}\delta_{\nu £} \quad (1.3)$$

where $\delta_{ƙ€}$ *and* $\delta_{\nu £}$ are the Krönecker delta defined as

$$\delta_{ƙ€} = \begin{cases} 0, & ƙ \neq € \\ 1, & ƙ = € \end{cases} \quad (1.4)$$

and

$$\delta_{\nu £} = \begin{cases} 0, & \nu \neq £ \\ 1, & \nu = £ \end{cases} \quad (1.5)$$

1.3.2 Polar Walsh Functions

Binary polar Walsh functions are derived from two dimensional Walsh functions expressed in polar coordinates and the orthogonality condition can be expressed as,

$$\frac{1}{\pi} \int_{r=0}^{1} \int_{\theta=0}^{2\pi} W_{ƙ}(r)\, W_{\nu}(\theta) W_{€}(r) W_{£}(\theta)\, rdr\, d\theta = \delta_{ƙ€}\delta_{\nu £} \quad (1.6)$$

For radial invariant case, the Exp. (1.6) reduces to

$$\int_0^{2\pi} W_{\nu}(\theta) W_{£}(\theta)\, d\theta \;=\; \frac{1}{2}\, \delta_{\nu £} \quad (1.7)$$

where $W_{\nu}(\theta)$ may be defined as azimuthal Walsh function of order ν.

A few two-dimensional Walsh functions is illustrated in Fig. 1.2 in polar coordinates.

1.4 Radial Walsh Functions

Radial Walsh functions [10] $W_{ƙ}(r)$ of index $ƙ \geq 0$ and argument r over the interval (0, 1) require integer $ƙ$ to be expressed in the form

$$ƙ \;=\; \sum_{i=0}^{\beta-1} ƙ_i\, 2^i \quad (1.8)$$

where $ƙ_i$ are the bits, 0 or 1 of the binary number $ƙ$ and 2^{β} is the integral power of 2 that just exceeds $ƙ$. Thus, ∀ r in (0, 1), we define,

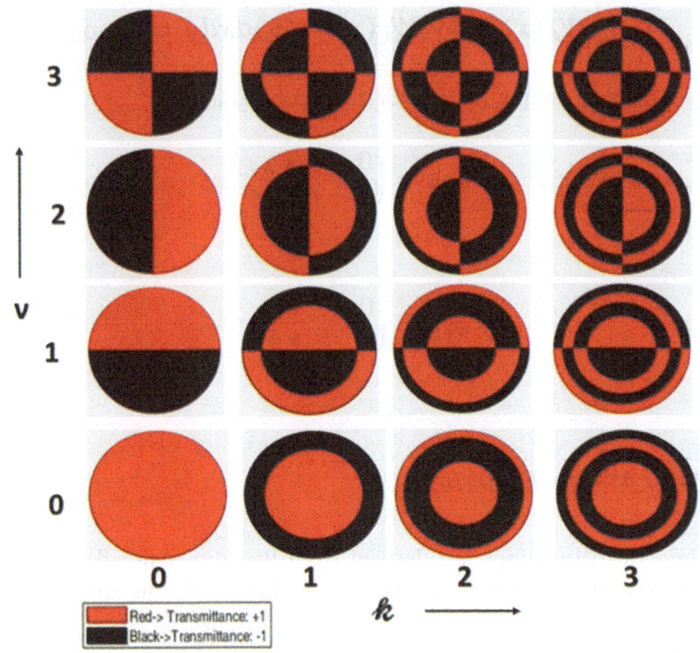

Fig. 1.2 Two dimensional Walsh functions in polar coordinates, $W_{k}(r)W_{\nu}(\theta)$. Red areas represent the values of transmission as +1 and black areas represent +1

$$W_{k}(r) = \prod_{i=0}^{\beta-1} sgn\left[\cos\left(2^{i}\pi t k_{i}\right)\right] = W_{k}(t) \tag{1.9}$$

where $t = r^2$. The function $sgn(x)$ is defined as,

$$\begin{aligned} sgn(x) &= +1, \quad x > 0 \\ &= 0, \quad x = 0 \\ &= -1, \quad x = 0 \end{aligned} \tag{1.10}$$

Figure 1.3 shows some radial Walsh functions generated in polar coordinates, in r space and the order k is equal to the number of zero crossings or the change in sign of the function in the interval (0, 1).

1.5 Azimuthal Walsh Functions

To define azimuthal Walsh function $W_{\nu}(\theta)$ of index $\nu \geq 0$ and argument θ over a sector bounded by 0 and 1 as inner and outer radii respectively, it is necessary to express the integer ν in the form,

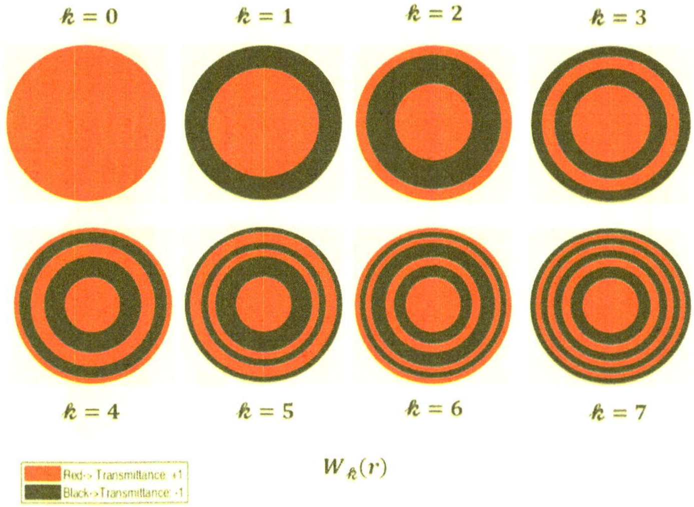

Fig. 1.3 Radial Walsh functions in r space, $W_{\mathcal{k}}(r)$, of orders $\mathcal{k} = 0, 1, 2, 3, 4, 5, 6, 7$.

$$\nu = \sum_{m=0}^{\alpha-1} \nu_m 2^m \tag{1.11}$$

where ν_m are the bits, 0 or 1 of the binary numerals for ν, and 2^α is the largest power of 2 that just exceeds ν. Thus, $\forall\theta$ in $(0, 2\pi)$, it can be defined as:

$$W_\nu(\theta) = \prod_{m=0}^{\alpha-1} sgn\left[\cos\left(\nu_m 2^m \frac{\theta}{2}\right)\right] \tag{1.12}$$

where the function $sgn(x)$ is defined in Exp. (1.10).

The orthogonality condition implies

$$\int_0^{2\pi} W_\nu(\theta) W_£(\theta)\, d\theta = \frac{1}{2}\,\delta_{\nu£} \tag{1.13}$$

where $\delta_{\nu£}$ is the Krönecker delta as defined in Exp. (1.5).

Figure 1.4 shows polar plots of azimuthal Walsh functions of orders 0, 1, 2 and 3.

Manipulating the values of ν_m azimuthal Walsh functions of orders from zero to sixty-three orders have been generated and shown in Table 1.1.

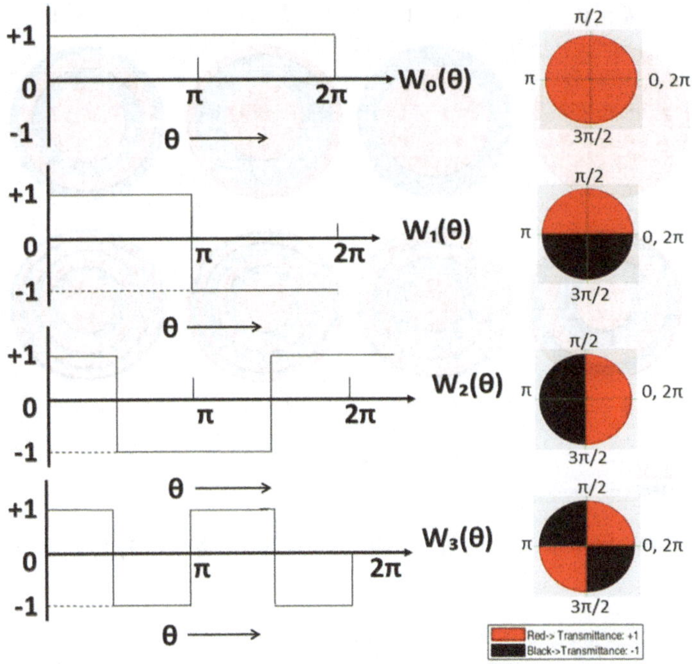

Fig. 1.4 Polar plot of Azimuthal Walsh functions $W_\nu(\theta)$, $\nu = 0, 1, 2, 3$

1.6 Self-Similarity in Azimuthal Walsh Functions

Azimuthal Walsh filters of orders within the range (0, 63) have been studied for finding out self-similar groups and sub-groups between different orders to examine self-similarity existing within them. Table 1.2 shows a vivid representation of azimuthal Walsh functions $W_\nu(\theta)$, for order $\nu = 0, 1, 2, \ldots, 63$ where Each two valued azimuthal Walsh function is represented by a row consisting of elements comprised of either '+' or '–' representing the values of $+1$ or -1 respectively corresponding to 0 or π phase differences. Each azimuthal Walsh function has been divided into equal parts within the domain (0, 2π) depending on the order of the Walsh function. For zeroth order azimuthal Walsh function $W_0(\theta)$, the value of the function is $+1$ over the entire domain (0, 2π). The minimum number of equal sectors, M required to define other orders of azimuthal Walsh functions is governed by the relationship $M = 2^\alpha$, where α is a positive integer satisfies the relation $2^{\alpha-1} \leq \nu \leq 2^\alpha$.

To obtain the higher orders of azimuthal Walsh functions, the technique proposed by Hazra and Mukherjee [10] is used whereas the unique rotational self-similarity between specific orders has also been observed. From $W_0(\theta)$, $W_1(\theta)$ can be derived by dividing the interval (0, 2π) in two equal sectors: the first sector is within (0, π) over which the phase is 0 and the second sector is within (π, 2π) over which the phase is π. Thus writing '+' in the left half and '–' in the right half. $W_2(\theta)$ can be similarly obtained from $W_1(\theta)$ by dividing each two sub-sectors of $W_1(\theta)$, namely

(0, π) and (π, 2π) into two equal sectors each, bounded by arcs of sector angles lying between 0 to π/2, π/2 to π and π to 3π/2, 3π/2 to 2 π respectively. Thus, four

Table 1.1 Orders and corresponding diagrams of Azimuthal Walsh functions $W_\nu(\theta)$ generated for $\nu = 0, 1, 2, \ldots, 63$

Order	Diagram	Order	Diagram	Order	Diagram
0		1		2	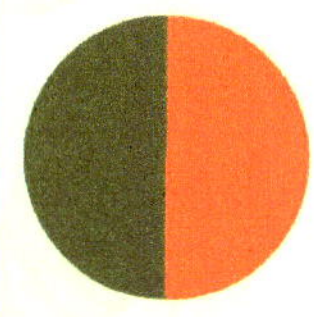
3	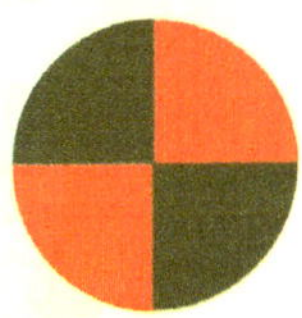	4		5	
6		7	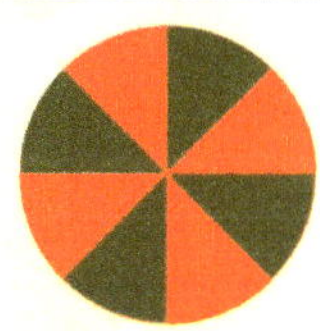	8	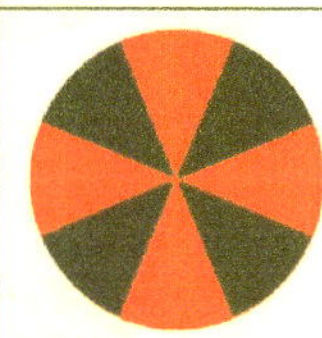
9	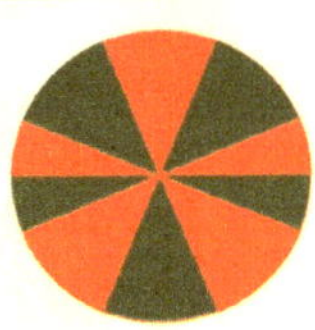	10	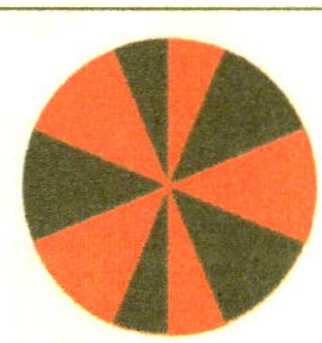	11	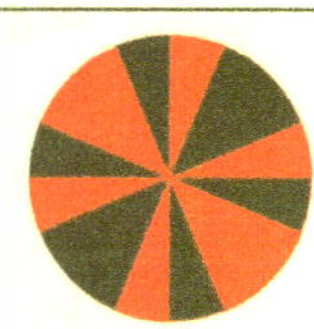
12		13		14	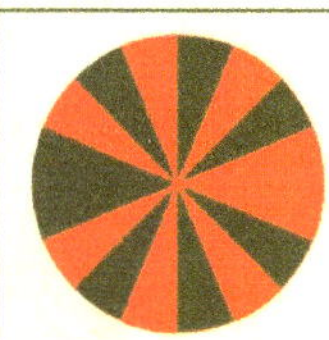
15		16		17	

Red-> Transmittance +1
Black->Transmittance -1

(continued)

Table 1.1 (continued)

Order	Diagram	Order	Diagram	Order	Diagram
18		19		20	
21		22		23	
24		25		26	
27		28		29	
30		31		32	
33		34		35	

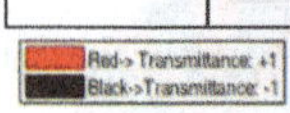

(continued)

Table 1.1 (continued)

Order	Diagram	Order	Diagram	Order	Diagram
36		37		38	
39		40		41	
42		43	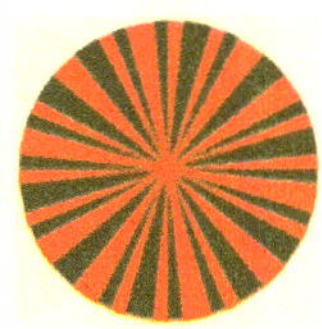	44	
45		46		47	
48		49	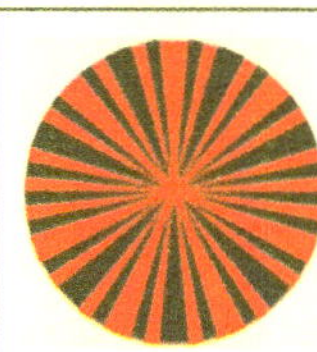	50	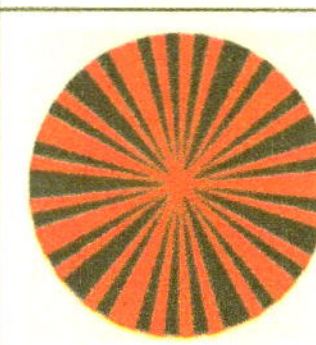
51	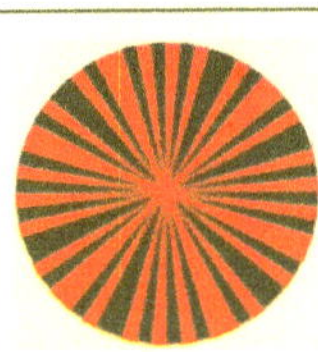	52		53	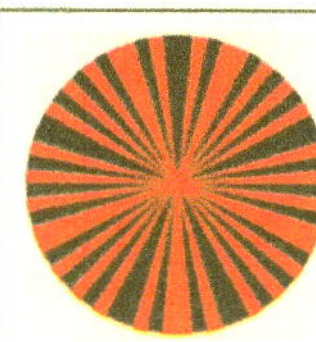

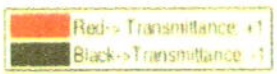

(continued)

Table 1.1 (continued)

Order	Diagram	Order	Diagram	Order	Diagram
54		55	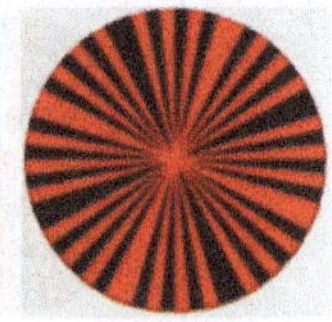	56	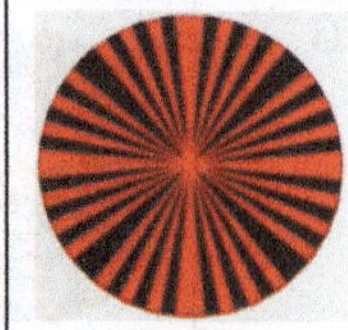
57	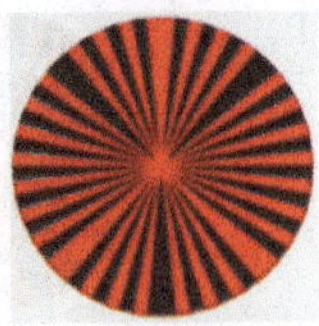	58	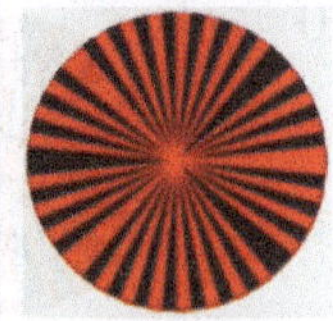	59	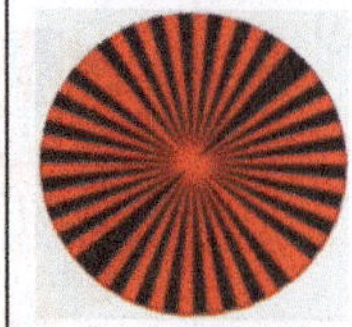
60	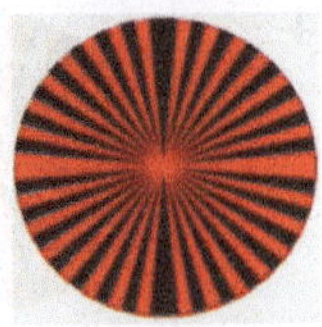	61	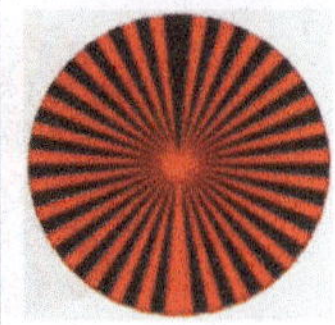	62	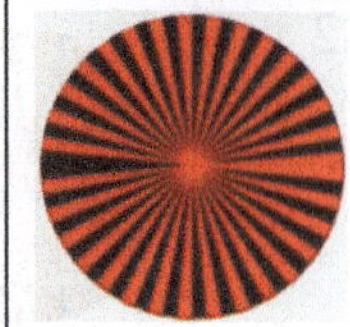
63	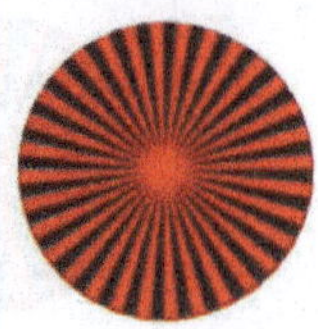				

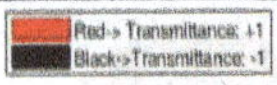

equal sectors are obtained and we write + − in the sub sectors where it was + before and − + where it was − before. It is observed that $W_2(\theta)$ is obtained from rotating $W_1(\theta)$ by $+\pi/2$ in the clockwise direction. Proceeding similarly, $W_3(\theta)$, consisting of four equal sectors of angles lying between 0 to $\pi/2$, $\pi/2$ to π, π to $3\pi/2$ and $3\pi/2$ to 2π, can be derived from $W_0(\theta)$ by + − wherever the value was + in $W_0(\theta)$.

$W_4(\theta)$ is derived from $W_3(\theta)$ and the angle of rotation is $+\ \pi/4$ of order $W_3(\theta)$ along the clockwise direction. $W_5(\theta)$ is derived from $W_4(\theta)$. $W_6(\theta)$ is derived from $W_1(\theta)$ and the angle of rotation in this case is $+3\pi/2$ in the clockwise direction. For a Walsh function of a higher order, choice of appropriate lower order from which it can be derived is decided by the requirement that order of the function is equal to the number of zero crossings. In the Table 1.2, the number on the right of each row gives the order of the azimuthal Walsh function corresponding to the row, and the number to the left of each row gives the order from which the function is originated.

Table 1.2 shows derivation of two valued azimuthal Walsh functions W_ν over the domain $(0, 2\pi)$ for $\nu = 0, 1, \ldots, 30$. Each two valued Walsh function is represented by a row. Elements of a row is denoted by either '+' ($\equiv +1$) or '−' ($\equiv -1$), provide values of the function over the sub intervals. The number on the last but one column gives the order of the Walsh function, and the number on the left gives the originating order of the Walsh function from it has been derived by means of alternating process. The last column on the right gives the angle rotation by which the previous order is to be rotated to find the present order azimuthal Walsh function.

Upon examining the set of azimuthal Walsh functions, it becomes apparent that the whole set of azimuthal Walsh functions does not comprise of self-similar structures. Distinct groups of Walsh functions exhibit self-similarity among their individual constituents. Group I consists of azimuthal Walsh functions 1, 3, 7, 15, 31, 63 ... all of which are derived from $W_0(\theta)$ and they are the primary group members. Members of Group II are derived from Group I. Similarly, members of Group III are derived from Group II and so on. Each group acts as a parent for one or more subgroups. Group I is unique. Members of Group I give rise to several subgroups namely IIA, II B, IIC, IID and so on. Each of the subgroups give rise to further subgroups. Taken together all of them constitute Group III whose individual subgroups are denoted by terms IIIAA, IIIAB, IIIAC, IIIBA, IIIBB, IIICA etc. In this nomenclature, the Roman numerals indicate the level of synthesis, followed by one or more letters of the alphabet to indicate the specific subgroup of the parent group from which it is generated; the last letter indicates the location of the first member of that specific subgroup with reference to the first member of other subgroups of the same level in an ascending order of sequence.

Table 1.2 Derivation of two-valued azimuthal Walsh functions W_ν over the domain $(0, 2\pi)$ for $\nu = 0, 1, \ldots, 30$

<table>
<tr><td>Origin</td><td colspan="32"></td><td>Order</td><td>Angle Rotation</td></tr>
<tr><td></td><td colspan="32">+</td><td>0</td><td>-</td></tr>
<tr><td></td><td colspan="34"></td></tr>
<tr><td>0</td><td colspan="16">+</td><td colspan="16">-</td><td>1</td><td>-</td></tr>
<tr><td></td><td colspan="34"></td></tr>
<tr><td>1</td><td colspan="8">+</td><td colspan="8">-</td><td colspan="8">-</td><td colspan="8">+</td><td>2</td><td>+π/2 of order 1</td></tr>
<tr><td>0</td><td colspan="8">+</td><td colspan="8">-</td><td colspan="8">+</td><td colspan="8">-</td><td>3</td><td>-</td></tr>
<tr><td></td><td colspan="34"></td></tr>
<tr><td>3</td><td colspan="4">+</td><td colspan="4">-</td><td colspan="4">-</td><td colspan="4">+</td><td colspan="4">+</td><td colspan="4">-</td><td colspan="4">-</td><td colspan="4">+</td><td>4</td><td>+π/4 of order 3</td></tr>
<tr><td>2</td><td colspan="4">+</td><td colspan="4">-</td><td colspan="4">-</td><td colspan="4">+</td><td colspan="4">-</td><td colspan="4">+</td><td colspan="4">+</td><td colspan="4">-</td><td>5</td><td>-</td></tr>
<tr><td>1</td><td colspan="4">+</td><td colspan="4">-</td><td colspan="4">+</td><td colspan="4">-</td><td colspan="4">-</td><td colspan="4">+</td><td colspan="4">-</td><td colspan="4">+</td><td>6</td><td>+3π/2 of order 5</td></tr>
<tr><td>0</td><td colspan="4">+</td><td colspan="4">-</td><td colspan="4">+</td><td colspan="4">-</td><td colspan="4">+</td><td colspan="4">-</td><td colspan="4">+</td><td colspan="4">-</td><td>7</td><td>-</td></tr>
<tr><td></td><td colspan="34"></td></tr>
<tr><td>7</td><td colspan="2">+</td><td colspan="2">-</td><td colspan="2">-</td><td colspan="2">+</td><td colspan="2">+</td><td colspan="2">-</td><td colspan="2">-</td><td colspan="2">+</td><td colspan="2">+</td><td colspan="2">-</td><td colspan="2">-</td><td colspan="2">+</td><td colspan="2">+</td><td colspan="2">-</td><td colspan="2">-</td><td colspan="2">+</td><td>8</td><td>+π/8 of order 7</td></tr>
<tr><td>6</td><td colspan="2">+</td><td colspan="2">-</td><td colspan="2">-</td><td colspan="2">+</td><td colspan="2">+</td><td colspan="2">-</td><td colspan="2">-</td><td colspan="2">+</td><td colspan="2">-</td><td colspan="2">+</td><td colspan="2">+</td><td colspan="2">-</td><td colspan="2">-</td><td colspan="2">+</td><td colspan="2">+</td><td colspan="2">-</td><td>9</td><td>-</td></tr>
<tr><td>5</td><td colspan="2">+</td><td colspan="2">-</td><td colspan="2">-</td><td colspan="2">+</td><td colspan="2">-</td><td colspan="2">+</td><td colspan="2">+</td><td colspan="2">-</td><td colspan="2">-</td><td colspan="2">+</td><td colspan="2">+</td><td colspan="2">-</td><td colspan="2">+</td><td colspan="2">-</td><td colspan="2">-</td><td colspan="2">+</td><td>10</td><td>+π/2 of order 9</td></tr>
<tr><td>4</td><td colspan="2">+</td><td colspan="2">-</td><td colspan="2">-</td><td colspan="2">+</td><td colspan="2">-</td><td colspan="2">+</td><td colspan="2">+</td><td colspan="2">-</td><td colspan="2">+</td><td colspan="2">-</td><td colspan="2">-</td><td colspan="2">+</td><td colspan="2">-</td><td colspan="2">+</td><td colspan="2">+</td><td colspan="2">-</td><td>11</td><td>-</td></tr>
<tr><td>3</td><td colspan="2">+</td><td colspan="2">-</td><td colspan="2">+</td><td colspan="2">-</td><td colspan="2">-</td><td colspan="2">+</td><td colspan="2">-</td><td colspan="2">+</td><td colspan="2">+</td><td colspan="2">-</td><td colspan="2">+</td><td colspan="2">-</td><td colspan="2">-</td><td colspan="2">+</td><td colspan="2">-</td><td colspan="2">+</td><td>12</td><td>+3π/4 of order 11</td></tr>
<tr><td>2</td><td colspan="2">+</td><td colspan="2">-</td><td colspan="2">+</td><td colspan="2">-</td><td colspan="2">-</td><td colspan="2">+</td><td colspan="2">-</td><td colspan="2">+</td><td colspan="2">-</td><td colspan="2">+</td><td colspan="2">-</td><td colspan="2">+</td><td colspan="2">+</td><td colspan="2">-</td><td colspan="2">+</td><td colspan="2">-</td><td>13</td><td>-</td></tr>
<tr><td>1</td><td colspan="2">+</td><td colspan="2">-</td><td colspan="2">+</td><td colspan="2">-</td><td colspan="2">+</td><td colspan="2">-</td><td colspan="2">+</td><td colspan="2">-</td><td colspan="2">-</td><td colspan="2">+</td><td colspan="2">-</td><td colspan="2">+</td><td colspan="2">-</td><td colspan="2">+</td><td colspan="2">-</td><td colspan="2">+</td><td>14</td><td>+π/2 of order 13</td></tr>
<tr><td>0</td><td colspan="2">+</td><td colspan="2">-</td><td colspan="2">+</td><td colspan="2">-</td><td colspan="2">+</td><td colspan="2">-</td><td colspan="2">+</td><td colspan="2">-</td><td colspan="2">+</td><td colspan="2">-</td><td colspan="2">+</td><td colspan="2">-</td><td colspan="2">+</td><td colspan="2">-</td><td colspan="2">+</td><td colspan="2">-</td><td>15</td><td>-</td></tr>
<tr><td></td><td colspan="34"></td></tr>
<tr><td>15</td><td>+</td><td>-</td><td>-</td><td>+</td><td>+</td><td>-</td><td>-</td><td>+</td><td>+</td><td>-</td><td>-</td><td>+</td><td>+</td><td>-</td><td>-</td><td>+</td><td>+</td><td>-</td><td>-</td><td>+</td><td>+</td><td>-</td><td>-</td><td>+</td><td>+</td><td>-</td><td>-</td><td>+</td><td>+</td><td>-</td><td>-</td><td>+</td><td>16</td><td>+π/16 of order 15</td></tr>
<tr><td>14</td><td>+</td><td>-</td><td>-</td><td>+</td><td>+</td><td>-</td><td>-</td><td>+</td><td>+</td><td>-</td><td>-</td><td>+</td><td>+</td><td>-</td><td>-</td><td>+</td><td>-</td><td>+</td><td>+</td><td>-</td><td>-</td><td>+</td><td>+</td><td>-</td><td>-</td><td>+</td><td>+</td><td>-</td><td>-</td><td>+</td><td>+</td><td>-</td><td>17</td><td>-</td></tr>
<tr><td>13</td><td>+</td><td>-</td><td>-</td><td>+</td><td>+</td><td>-</td><td>-</td><td>+</td><td>-</td><td>+</td><td>+</td><td>-</td><td>-</td><td>+</td><td>+</td><td>-</td><td>-</td><td>+</td><td>+</td><td>-</td><td>-</td><td>+</td><td>+</td><td>-</td><td>+</td><td>-</td><td>-</td><td>+</td><td>+</td><td>-</td><td>-</td><td>+</td><td>18</td><td>+π/2 of order 17</td></tr>
<tr><td>12</td><td>+</td><td>-</td><td>-</td><td>+</td><td>+</td><td>-</td><td>-</td><td>+</td><td>-</td><td>+</td><td>+</td><td>-</td><td>-</td><td>+</td><td>+</td><td>-</td><td>+</td><td>-</td><td>-</td><td>+</td><td>+</td><td>-</td><td>-</td><td>+</td><td>-</td><td>+</td><td>+</td><td>-</td><td>-</td><td>+</td><td>+</td><td>-</td><td>19</td><td>-</td></tr>
<tr><td>11</td><td>+</td><td>-</td><td>-</td><td>+</td><td>-</td><td>+</td><td>+</td><td>-</td><td>-</td><td>+</td><td>+</td><td>-</td><td>+</td><td>-</td><td>-</td><td>+</td><td>+</td><td>-</td><td>-</td><td>+</td><td>-</td><td>+</td><td>+</td><td>-</td><td>-</td><td>+</td><td>+</td><td>-</td><td>+</td><td>-</td><td>-</td><td>+</td><td>20</td><td>+3π/4 of order 19</td></tr>
<tr><td>10</td><td>+</td><td>-</td><td>-</td><td>+</td><td>-</td><td>+</td><td>+</td><td>-</td><td>-</td><td>+</td><td>+</td><td>-</td><td>+</td><td>-</td><td>-</td><td>+</td><td>-</td><td>+</td><td>+</td><td>-</td><td>+</td><td>-</td><td>-</td><td>+</td><td>+</td><td>-</td><td>-</td><td>+</td><td>-</td><td>+</td><td>+</td><td>-</td><td>21</td><td>-</td></tr>
<tr><td>9</td><td>+</td><td>-</td><td>-</td><td>+</td><td>-</td><td>+</td><td>+</td><td>-</td><td>+</td><td>-</td><td>-</td><td>+</td><td>-</td><td>+</td><td>+</td><td>-</td><td>-</td><td>+</td><td>+</td><td>-</td><td>+</td><td>-</td><td>-</td><td>+</td><td>-</td><td>+</td><td>+</td><td>-</td><td>+</td><td>-</td><td>-</td><td>+</td><td>22</td><td>+π/2 of order 21</td></tr>
<tr><td>8</td><td>+</td><td>-</td><td>-</td><td>+</td><td>-</td><td>+</td><td>+</td><td>-</td><td>+</td><td>-</td><td>-</td><td>+</td><td>-</td><td>+</td><td>+</td><td>+</td><td>+</td><td>-</td><td>-</td><td>+</td><td>-</td><td>+</td><td>+</td><td>-</td><td>+</td><td>-</td><td>-</td><td>+</td><td>-</td><td>+</td><td>+</td><td>-</td><td>23</td><td>-</td></tr>
<tr><td>7</td><td>+</td><td>-</td><td>+</td><td>-</td><td>-</td><td>+</td><td>-</td><td>+</td><td>+</td><td>-</td><td>+</td><td>-</td><td>-</td><td>+</td><td>-</td><td>+</td><td>+</td><td>-</td><td>+</td><td>-</td><td>-</td><td>+</td><td>-</td><td>+</td><td>+</td><td>-</td><td>+</td><td>-</td><td>-</td><td>+</td><td>-</td><td>+</td><td>24</td><td>+π/16 of order 23</td></tr>
<tr><td>6</td><td>+</td><td>-</td><td>+</td><td>-</td><td>-</td><td>+</td><td>-</td><td>+</td><td>+</td><td>-</td><td>+</td><td>-</td><td>-</td><td>+</td><td>-</td><td>-</td><td>-</td><td>+</td><td>-</td><td>+</td><td>+</td><td>-</td><td>+</td><td>-</td><td>-</td><td>+</td><td>-</td><td>+</td><td>+</td><td>-</td><td>+</td><td>-</td><td>25</td><td>-</td></tr>
<tr><td>5</td><td>+</td><td>-</td><td>+</td><td>-</td><td>-</td><td>+</td><td>-</td><td>+</td><td>-</td><td>+</td><td>-</td><td>+</td><td>+</td><td>-</td><td>+</td><td>-</td><td>-</td><td>+</td><td>-</td><td>+</td><td>+</td><td>-</td><td>+</td><td>-</td><td>+</td><td>-</td><td>+</td><td>-</td><td>-</td><td>+</td><td>-</td><td>+</td><td>26</td><td>+π/2 of order 25</td></tr>
<tr><td>4</td><td>+</td><td>-</td><td>+</td><td>-</td><td>-</td><td>+</td><td>-</td><td>+</td><td>-</td><td>+</td><td>-</td><td>+</td><td>+</td><td>-</td><td>+</td><td></td><td>+</td><td>-</td><td>+</td><td>-</td><td>-</td><td>+</td><td>-</td><td>+</td><td>-</td><td>+</td><td>-</td><td>+</td><td>+</td><td>-</td><td>+</td><td>-</td><td>27</td><td>-</td></tr>
<tr><td>3</td><td>+</td><td>-</td><td>+</td><td>-</td><td>+</td><td>-</td><td>+</td><td>-</td><td>-</td><td>+</td><td>-</td><td>+</td><td>-</td><td>+</td><td>-</td><td>+</td><td>+</td><td>-</td><td>+</td><td>-</td><td>+</td><td>-</td><td>+</td><td>-</td><td>-</td><td>+</td><td>-</td><td>+</td><td></td><td>+</td><td>-</td><td>+</td><td>28</td><td>+3π/4 of order 27</td></tr>
<tr><td>2</td><td>+</td><td>-</td><td>+</td><td>-</td><td>+</td><td>-</td><td>+</td><td>-</td><td>-</td><td>+</td><td>-</td><td>+</td><td>-</td><td>+</td><td>-</td><td>+</td><td>-</td><td>+</td><td>-</td><td>+</td><td>-</td><td>+</td><td>-</td><td>+</td><td>+</td><td>-</td><td>+</td><td>-</td><td>+</td><td>-</td><td>+</td><td>-</td><td>29</td><td>-</td></tr>
<tr><td>1</td><td>+</td><td>-</td><td>+</td><td>-</td><td>+</td><td>-</td><td>+</td><td>-</td><td>+</td><td>-</td><td>+</td><td>-</td><td>+</td><td>-</td><td>+</td><td>-</td><td>-</td><td>+</td><td>-</td><td>+</td><td>-</td><td>+</td><td>-</td><td>+</td><td>-</td><td>+</td><td>-</td><td>+</td><td>-</td><td>+</td><td>-</td><td>+</td><td>30</td><td>+π/2 of order 29</td></tr>
<tr><td>0</td><td>+</td><td>-</td><td>+</td><td>-</td><td>+</td><td>-</td><td>+</td><td>-</td><td>+</td><td>-</td><td>+</td><td>-</td><td>+</td><td>-</td><td>+</td><td>-</td><td>+</td><td>-</td><td>+</td><td>-</td><td>+</td><td>-</td><td>+</td><td>-</td><td>+</td><td>-</td><td>+</td><td>-</td><td>+</td><td>-</td><td>+</td><td>-</td><td>31</td><td>-</td></tr>
</table>

0 $\theta \longrightarrow$ 2π

The formations of groups and subgroups for the first few levels are illustrated in Tables 1.3, 1.4, 1.5, 1.6 and 1.7 where members of each group are given by the expression [8],

$$Z(\sigma) = 2^{\sigma+b} + c, \quad \sigma = 1, 2, 3, \ldots \tag{1.14}$$

b and c are two parameters specific for each subgroup where $b \geq 0$ and $c = -1, 0$ or dependent on σ.

Table 1.3 Composition of self-similar groups and subgroups of azimuthal Walsh functions (Group I and II A)

Group	Sub group	First few order members of the group					
I	-	1	3	7	15	31	
		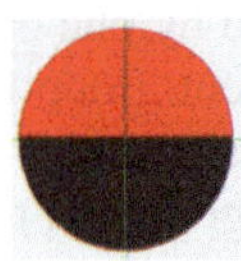	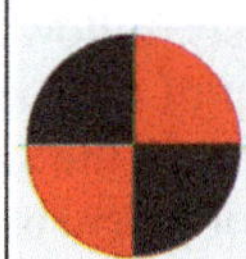	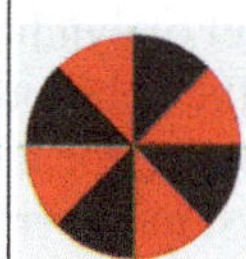	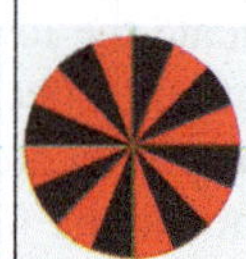		....
		$\sigma =$ 1	2	3	4	5	
		Originating Order					
		0 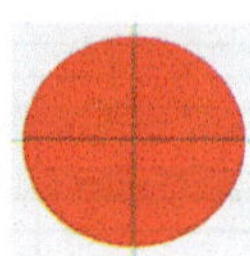					
		b	**c**		**$Z(\sigma)$**		
		0	**-1**		**2^{σ}-1**		
II	A	2	4	8	16	32	
		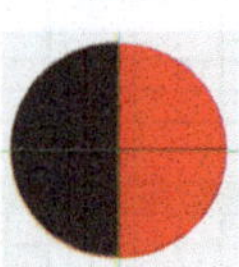					....
		Originating Orders					
		1	3	7	15	31	....
		b	**c**		**$Z(\sigma)$**		
		0	**0**		**2^{σ}**		

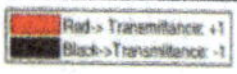

Table 1.4 Composition of self-similar groups and subgroups of azimuthal Walsh functions (Group II B, C, D)

Group	Sub group	First few order members of the group				
II	B	6 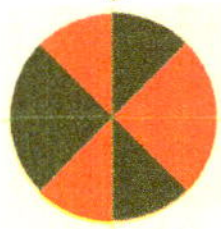 $\sigma = 1$	12  2	24  3	48  4	.....
		Originating Orders				
		1	3	7	15	.....
		b	**c**		**$Z(\sigma)$**	
		1	$\mathbf{2^{\sigma}}$		$\mathbf{2^{\sigma+1} + 2^{\sigma}}$	
	C	14 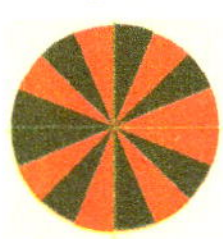	28 	56 	.....	
		Originating Orders				
		1	3	7	.....	
		b	**c**		**$Z(\sigma)$**	
		2	$\mathbf{2^{\sigma+1} + 2^{\sigma}}$		$\mathbf{2^{\sigma+2} + \left(2^{\sigma+1} + 2^{\sigma}\right)}$	
	D	30 	60 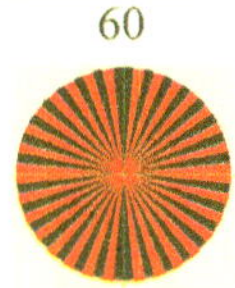	.....		
		Originating Orders				
		1	3	.....		
		b	**c**		**$Z(\sigma)$**	
		3	$\mathbf{2^{\sigma+2} + \left(2^{\sigma+1} + 2^{\sigma}\right)}$		$\mathbf{2^{\sigma+3} + \left(2^{\sigma+2} + \left(2^{\sigma+1} + 2^{\sigma}\right)\right)}$	
	...	...	...		...	

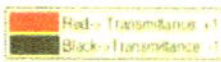

(continued)

Table 1.4 (continued)

Group	Sub group	First few order members of the group				
III	AA	5 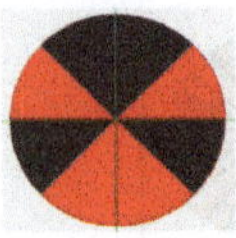 $\sigma = 1$	11 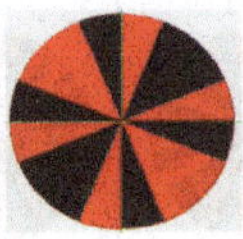 2	23  3	47 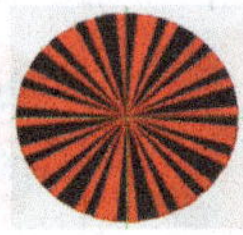 4	......
		Originating Orders				
		2	4	8	16	......
		b	**c**		$Z(\sigma)$	
		1	$2^{\sigma} - 1$		$2^{\sigma+1} + (2^{\sigma} - 1)$	
	AB	13 	27 	55 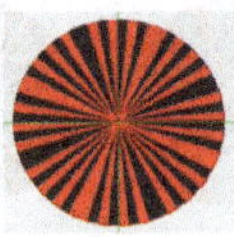	......	
		Originating Orders				
		2	4	8	......	
		b	**c**		$Z(\sigma)$	
		2	$2^{\sigma+1} + (2^{\sigma} - 1)$		$2^{\sigma+2} + \left(2^{\sigma+1} + (2^{\sigma} - 1)\right)$	
	AC	29 	59 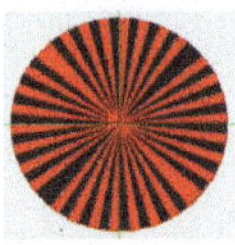	......		
		Originating Orders				
		2	4	......		
		b	**c**		$Z(\sigma)$	
		3	$2^{\sigma+2} + \left(2^{\sigma+1} + (2^{\sigma} - 1)\right)$		$2^{\sigma+3} + \left(2^{\sigma+2} + \left(2^{\sigma+1} + (2^{\sigma} - 1)\right)\right)$	
	...	...	...		...	

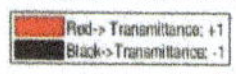

Table 1.5 Composition of self-similar groups and subgroups of azimuthal Walsh functions (Group III AA, AB, AC)

Group	Sub group	First few order members of the group			
III	BA	9 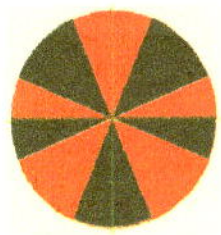 $\sigma = 1$	19  2	39  3	……
		Originating Orders			
		6	12	24	……
		b	**c**		$\boldsymbol{Z(\sigma)}$
		2	$\mathbf{2^{\sigma} - 1}$		$\mathbf{2^{\sigma+2} + (2^{\sigma} - 1)}$
	BB	25 	51 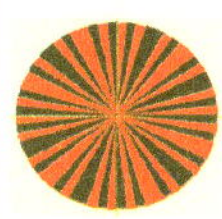	……	
		Originating Orders			
		6	12	……	
		b	**c**		$\boldsymbol{Z(\sigma)}$
		3	$\mathbf{2^{\sigma+2} + (2^{\sigma} - 1)}$		$\mathbf{2^{\sigma+3} + \left(2^{\sigma+2} + (2^{\sigma} - 1)\right)}$
	BC …..	57 	……		
		Originating Order			
		6	……		
		b	**c**		$\boldsymbol{Z(\sigma)}$
		4	$\mathbf{2^{\sigma+3} + \left(2^{\sigma+2} + (2^{\sigma} - 1)\right)}$		$\mathbf{2^{\sigma+4} + \left(2^{\sigma+3} + \left(2^{\sigma+2} + (2^{\sigma} - 1)\right)\right)}$
	…	…	…		…

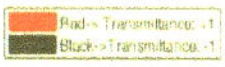

Table 1.6 Composition of self-similar groups and subgroups of azimuthal Walsh functions (Group III BA, BB, BC)

Group	Sub group	First few order members of the group			
III	CA	17  $\sigma = 1$	35 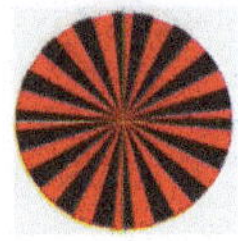 2	......	
		Originating Orders			
		14	28	......	
		b	**c**	$\boldsymbol{Z(\sigma)}$	
		3	$\mathbf{2^{\sigma}-1}$	$\mathbf{2^{\sigma+3}+(2^{\sigma}-1)}$	
	CB	49 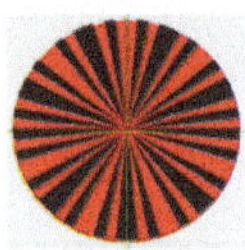	......		
		Originating Orders			
		14	......		
		b	**c**	$\boldsymbol{Z(\sigma)}$	
		4	$\mathbf{2^{\sigma+3}+(2^{\sigma}-1)}$	$\mathbf{2^{\sigma+4}+\left(2^{\sigma+3}+(2^{\sigma}-1)\right)}$	
	...	...	...	...	
IV	AAA	10 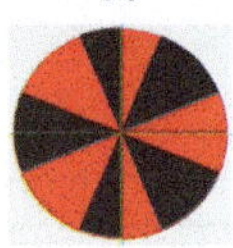 $\sigma = 1$	20  2	40  3	......
		Originating Orders			
		14	28	56	......
		b	**c**		$\boldsymbol{Z(\sigma)}$
	....	**2**	$\mathbf{2^{\sigma}}$		$\mathbf{2^{\sigma+2}+2^{\sigma}}$

Table 1.7 Composition of self-similar groups and subgroups of azimuthal Walsh functions (Group III CA, CB, Group IV AAA, AAB, BAA and Group V AAAA…)

IV	AAB	49 		……
		Originating Orders		
		14		……
		b	**c**	$\boldsymbol{Z(\sigma)}$
		4	$\mathbf{2^{\sigma+3}+(2^{\sigma}-1)}$	$\mathbf{2^{\sigma+4}+\left(2^{\sigma+3}+(2^{\sigma}-1)\right)}$
	…	…	…	…
IV	BAA	22  $\sigma=$ 1	44  2	……
		Originating Orders		
	….	9 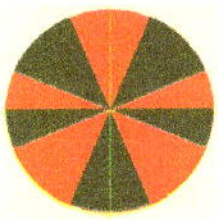	19 	……
		b	**c**	$\boldsymbol{Z(\sigma)}$
		2	$\mathbf{2^{\sigma}}$	$\mathbf{2^{\sigma+2}+2^{\sigma}}$
	….	….		
V	AAAA	21 	43 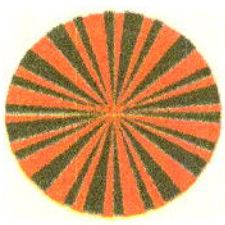	……
		Originating Orders		
		10	20	……
		b	**c**	$\boldsymbol{Z(\sigma)}$
		3	$\mathbf{2^{\sigma+2}-(2^{\sigma}+1)}$	$\mathbf{2^{\sigma+3}+[2^{\sigma+2}-(2^{\sigma}+1)]}$
	….	….	….	….

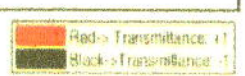

From the Tables it is apparent that not only higher order azimuthal Walsh functions can be derived from members of each group and subgroup of lower orders, but new subgroups will also emerge when Walsh functions of higher orders are taken into consideration. For the subgroups already identified, members lying in higher orders can be easily identified by using the analytical expression Eq. (1.14) for $Z(\sigma)$.

1.7 Walsh Filters

Walsh filters of various orders may be obtained from the corresponding Walsh functions by realizing the values of $+1$ or -1 with 0 or π phase respectively.

1.7.1 Radial Walsh Filters

Radial Walsh filters, derived from radial Walsh functions, form a set of orthogonal phase filters that takes on phase values of either 0 or π, corresponding to $+1$ or -1 values of the radial Walsh functions. Order of these filters is given by the number of zero crossings, or equivalently phase transitions within the domain over which the set is defined. In general, radial Walsh filters are binary phase filters or zone plates, each of them demonstrating distinct focusing characteristics.

1.7.2 Azimuthal Walsh Filters

Azimuthal Walsh filters are derived from azimuthal Walsh functions, also form a set of orthogonal phase filters that takes on values $+1$ or -1 of the azimuthal Walsh functions specifies corresponding phase values 0 or π. Order of these filters is given by the number of phase transitions within the angular domain of $(0, 2\pi)$. Azimuthal Walsh filter can be represented by sector shaped aperture bounded by constant inner and outer radii with varying angles θ_1 and θ_2 within the domain $(0, 2\pi)$. The values of inner and outer radii are taken as $R_1 = 0$ and $R_2 = 1$. In general, azimuthal Walsh filters can be considered as combination of binary phase filters or azimuthally varying zone plates which are having unique properties of introducing phase factors to the interacting light beam.

1.8 Self-Similarity in Azimuthal Walsh Filters

The set of azimuthal Walsh filters can be classified into distinct self-similar groups and sub-groups in the same manner as the azimuthal Walsh functions from which they have been derived [11]. Group I is unique. It has no subgroups. Members of this group possess self-similar structures as classified in Table 1.3. Groups IIA and above can similarly be divided into subgroups as represented in the Tables 1.4, 1.5, 1.6 and 1.7. Members of each subgroup possess distinct self-similar structures or phase sequences.

References

1. J.L. Walsh, A closed set of normal orthogonal functions. Am. J. Math. **45**(1), 5–24 (1923)
2. H.F. Harmuth, *Transmission of Information by Orthogonal Functions* (Springer, Berlin, 1972), p. 31
3. K.G. Beauchamp, *Walsh Functions and Their Applications* (Academic Press, New York, 1985)
4. H.C. Andrews, *Computer Techniques in Image Processing* (Academic, New York, 1970)
5. L.N. Hazra, A. Banerjee, Application of Walsh Function in Generation of Optimum Apodizers. J. Opt. (India) **5**, 19–26 (1976)
6. M. De, L.N. Hazra, Walsh functions in problems of optical imagery. Opt. Acta **24**, 221–234 (1977)
7. L.N. Hazra, A. Guha, Farfield diffraction properties of radial Walsh filters. J. Opt. Soc. Am. A **3**(6), 843–846 (1986)
8. L.N. Hazra, Walsh filters in tailoring of resolution in microscopic imaging. Micron **38**(2), 129–135 (2007)
9. B.B. Mandelbrot, *The Fractal Geometry of Nature* (Freeman, San Fransisco, California, 1982)
10. P. Mukherjee, L.N. Hazra, Self-similarity in radial Walsh filters and axial intensity distributions in the far-field diffraction pattern. J. Opt. Soc. Am. A **31**(2), 379–387 (2014)
11. I. Bhattacharya, in Self-similarity in azimuthal Walsh filters and corresponding far-field diffraction characteristics: a unique study to control tightly focused fields and coupling of light into metamaterials, plasmonic structure and waveguides, in *Proc. SPIE 11257, Plasmonics in Biology and Medicine* XVII:1125717 (2020), pp. 1–14

Chapter 2
Transverse Intensity Distribution on the Far-Field Plane of Azimuthal Walsh Filters

2.1 Introduction

Toraldo Di Francia [1–3], pioneered investigations on pupil plane filtering to enhance the resolving power of imaging system beyond the diffraction limit. Walsh filters derived from radial Walsh functions can be used as pupil plane filters for tailoring the resolution in microscopic imaging [4]. They are used effectively in tackling the problem of apodization [5–7], and in studies on aberrated optical imagery [8–10]. The far-field amplitude characteristics of some of these filters have been studied [11] and they show the potential for their effective use in various applications like high resolution microscopy, optical data storage, microlithography, optical encryption and optical micromanipulation [12–18]. There has been a keen interest in the imaging properties of sector shaped apertures and have been investigated by many researchers. These apertures play significant roles in many applications such as holography, space astronomy, micro-lithography etc., to name a few! The basic form of this type of aperture is the circular aperture, bounded by specific values of radius and azimuth and the far-field diffraction pattern of this aperture while illuminated by coherent beam of light is the well-known Airy pattern [19]. It has been assumed that the amplitude and phase remain constant over the aperture. The analysis of far-field diffraction characteristics [20] has been studied using the technique of calculating intensity point spread function for sector shaped apertures proposed by Mahan et al. [21]. Lessard and Som [22] has extended the idea both theoretically and practically and used this aperture for holographic multiplexing technique used by the authors. Azimuthal Walsh filters are derived from the azimuthal Walsh functions. Diffraction theory for azimuthally structured Fresnel Zone Plate has been studied by T. Vierke and Jürgen [23] where they have studied the light transmission from concentric rings with alternating binary transmission of zero and one and the said transmission has been modulated in the azimuthal direction. The resulting structure reported by them is

I. Bhattacharya and L. Hazra, *Azimuthal Walsh Filters*, Progress in Optical Science and Photonics 10,
https://doi.org/10.1007/978-981-15-6099-6_2

of importance in extreme UV and X-ray imaging and shows good focal properties that can be obtained with it. Practical realization of azimuthal Walsh filters is facilitated by the availability of high-speed Spatial Light modulators [24] where the in situ generation of different orders of azimuthal Walsh filters are possible.

In this chapter we present the far-field intensity distribution on the transverse image plane of a rotationally symmetric imaging system, with azimuthal Walsh filter of specific order placed on the exit pupil plane. Azimuthal Walsh filters are represented by a combination of sector-shaped apertures bounded by unity value of radius and specific values of azimuths. The zero order azimuthal Walsh filter corresponds to circular aperture of unity amplitude, and the corresponding far-field diffraction pattern is the well-known Airy pattern. As the order increases, the point spread functions show pronounced asymmetry introduced by the phases of azimuthal Walsh filters [25, 26]. The results of intensity distribution on the far-field plane are validated with theoretically known results for some pupils. Experimental verification using SLMs with analytically computed values has also been represented. The azimuthal Walsh filters can also be explored as phase-shifted zone plates to generate complex 3D beam structures like petal shaped and optical ring lattice beams [27].

2.2 Analytical Formulation of Far-Field Amplitude Distribution Along an Azimuth for a Single Sector on the Exit Pupil

Let us start with a single sector-shaped aperture, AE′B, subtending an angle ϕ at the centre E' on the exit pupil of radius R of a rotationally symmetric optical imaging system (Fig. 2.1).

Thus,

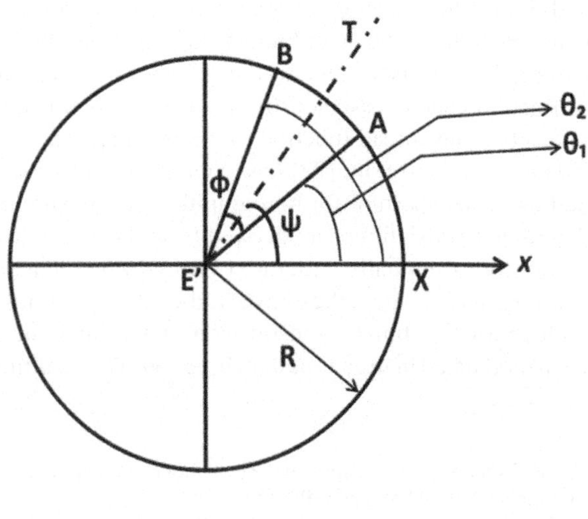

Fig. 2.1 Schematic diagram of a sector shaped aperture, AE′B of angle ϕ

$$\angle AE'B = \phi \tag{2.1}$$

Let E′T be the bisector of the aperture AE′B making an angle ψ with the x-axis along E′X so that:

$$\angle TE'X = \psi \tag{2.2}$$

Hence,

$$\angle AE'X = \theta_1 = \psi - \phi/2 \tag{2.3}$$

$$\angle BE'X = \theta_2 = \psi + \phi/2 \tag{2.4}$$

The amplitude transmittance, T on the sector is given by:

$$T = e^{i\chi} \tag{2.5}$$

where χ represents the phase factor.

Representing azimuthal Walsh functions, by single or combinations of sector-shaped apertures, the phase term becomes,

$$\chi = \pi \text{ or } -\pi \tag{2.6}$$

which implies the value of transmittance T,

$$T = +1 \text{ or } -1 \tag{2.7}$$

on different sectors. The schematic diagram of a rotationally symmetric imaging system with azimuthal Walsh filter placed at the exit pupil plane is shown in Fig. 2.2.

The normalized coordinates for a point, P on the pupil is (r, θ), where $r = \rho/R$, ρ is the radial distance of P from the centre of the exit pupil, and R is the radius of the pupil as depicted in Fig. 2.3.

Normalized coordinates of a point Q on the far-field plane is defined as (p, ζ) where

$$p = \frac{2\pi}{\lambda} \cdot \frac{R}{f} \cdot \xi \tag{2.8}$$

p is called the reduced diffraction variable at far-field plane, $\frac{2\pi}{\lambda}$ is the propagation constant, f being the focal length of the imaging system used, and $\xi(= O'Q)$ is the geometrical distance of the point of observation on the far-field plane with respect to the centre O′ of the diffraction pattern.

Exp. (2.8) can be rewritten in terms of numerical aperture of the pupil plane as,

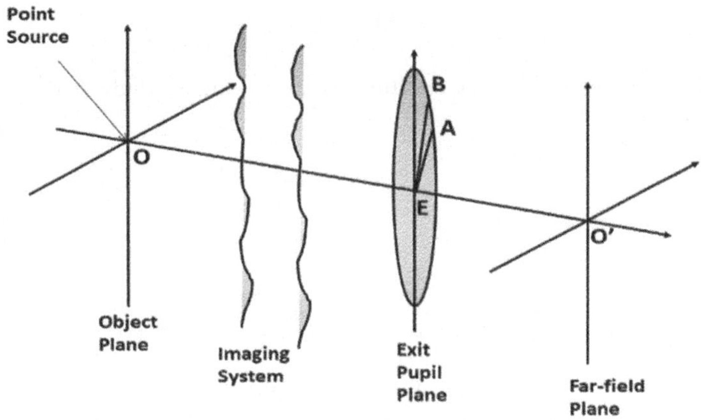

Fig. 2.2 Schematic diagram of a rotationally symmetric imaging system

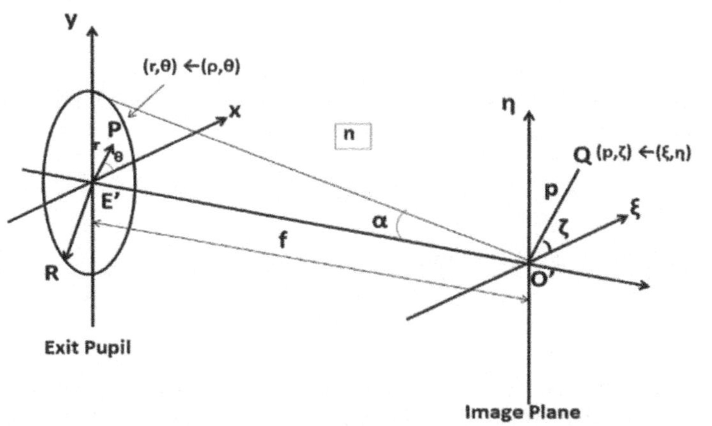

Fig. 2.3 A point *P* on the exit pupil plane and a point Q on a transverse imaging plane in the far-field

$$p = \frac{2\pi}{\lambda} \cdot (n \sin \alpha) \cdot \xi \tag{2.9}$$

where α is the angle subtended by the pupil plane at the centre of the far-field plane and $(n \sin \alpha)$ is the image space numerical aperture, n is the refractive index of the image space and λ is the operating wavelength.

Following the complex notation as suggested by Born [1], complex amplitude on the far-field plane can be expressed along with a multiplicative constant:

$$A\,(p,\zeta) = \int_0^R \int_0^{2\pi} f(r,\theta)\, exp[i\, p\, r \cos\omega]\, r\, dr\, d\theta \tag{2.10}$$

where $\omega = (\theta - \zeta)$.

2.3 Azimuthal Walsh Filters on the Exit Pupil

In case of radially invariant filters, the pupil function or transmittance $f(r, \theta)$ of a sector shaped aperture with sector angle extending from θ_{m-1} to θ_m reduces to $f(\theta)$ and is expressed as:

$$\begin{aligned} f(r,\theta) &= f(\theta), \text{ over the aperture, } \theta_{m-1} \le \theta \le \theta_m \\ &= 0, \quad \text{elsewhere} \end{aligned} \tag{2.11}$$

Representing azimuthally variant Walsh filter by $W_\nu(\theta)$ of order ν on the pupil, the pupil function may be expressed as:

$$f(\theta) = W_\nu(\theta) = \sum_{m=1}^{M} e^{ik\psi_m}\, \mathfrak{B}_m(\theta) \tag{2.12}$$

where $\mathfrak{B}_m(\theta)$ are azimuthal Walsh Block functions defined as:

$$\begin{aligned} \mathfrak{B}_m(\theta) &= 1; \quad \theta_{m-1} \le \theta \le \theta_m \\ &= 0; \quad \text{otherwise} \end{aligned} \tag{2.13}$$

where

$$\theta_{m-1} = \left(\frac{m-1}{M}\right)2\pi \text{ and } \theta_m = \left(\frac{m}{M}\right)2\pi \tag{2.14}$$

with $M = 2^\alpha$ is the largest power of 2 that just exceeds ν. Over the mth sector of $W_\nu(\theta)$, value of $k\psi_m$ is 0 or π, if $W_\nu(\theta) = +1$ or $-1 = +1$ or $-$ respectively.

Substituting in Eq. (2.10), we can write:

$$A(p,\zeta) = \int_0^1 \int_0^{2\pi} \sum_{m=1}^{M} e^{ik\psi_m} \mathfrak{B}_m(\theta) \exp[ipr\cos\omega] r \mathrm{d}r \mathrm{d}\theta \tag{2.15}$$

Evaluation of the integration in (3.14) involving the exponential has been done as suggested by Mahan et al. [2], and Som and Lessard [3], and the complex amplitude distribution $A(p, \zeta)$ can be expressed as:

$$A(p,\zeta) = T[C(p,\zeta;\theta) - iS(p,\zeta;\theta)] \tag{2.16}$$

where T is the transmission factor of the pupil function given by:

$$T = \sum_{m=1}^{M} e^{ik\psi_m} \tag{2.17}$$

and $C(p, \zeta; \theta)$ and $S(p, \zeta; \theta)$ are the real and imaginary parts of $A(p, \zeta)$ at the far-field plane can be expressed as the series of Bessel functions of even and odd arguments as:

$$C(p, \zeta; \theta) = \phi \frac{J_1(p)}{p} + \left\{ \sum_{n=1}^{\infty} \frac{2n+1}{n(n+1)} \frac{J_{2n+1}(p)}{p} X \sum_{m=1}^{n} [\sin 2m(\theta_2 - \zeta) - \sin 2m(\theta_1 - \zeta)] \right\} \tag{2.18}$$

and

$$S(p, \zeta; \theta) = 8 \sum_{n=1}^{\infty} \frac{n}{4n^2 - 1} \frac{J_{2n}(p)}{p} \sum_{m=1}^{n} (-1)^{m-1} [\sin(2m-1)(\theta_2 - \zeta) - \sin(2m-1)(\theta_1 - \zeta)] \tag{2.19}$$

respectively.

Each azimuthal Walsh function, $W_\nu(\theta)$, $\nu = 1, 2, 3, \ldots$ may be considered consisting of 2^N no. of sectors where the value of N depends on the order ν. For example, for W_0, W_1, W_2 and W_2, $N = 4$. Value of transmission T in each of the four sectors is either $+1$ or -1, as per the order of the Walsh function.

Far-field amplitude distribution for each of these Walsh functions, $W_\nu(\theta)$ can be obtained as:

$$A(p, \zeta) = \sum_{s=0}^{3} T_s[C_s(p, \zeta; \theta) - iS_s(p, \zeta; \theta)] \tag{2.20}$$

Corresponding intensity distribution, $I(p, \zeta)$ can be calculated as

$$I(p, \zeta) = \left[\sum_{s=0}^{3} T_s[C_s(p, \zeta; \theta)] \right]^2 + \left[\sum_{s=0}^{3} T_s[S_s(p, \zeta; \theta)] \right]^2 \tag{2.21}$$

The normalized intensity point spread function at the far-field plane $I_N(p, \zeta)$ is given by,

$$I_N(p, \zeta) = I(p, \zeta)/I(0, 0) \tag{2.22}$$

where $I(0, 0)$ is the intesity distribution at the origin of the far-field plane where $p = 0, \zeta = 0$.

2.4 Asymmetrical Amplitude Point Spread Function on the Far-Field Plane Due to Azimuthal Walsh Filter at Exit Pupil Plane

The azimuthal Walsh filter $W_\upsilon(\theta)$ of order $\upsilon = 1, 2, 3, \ldots$ is placed on the exit pupil of a rotationally symmetric imaging system (Fig. 2.2) and the complex amplitude distribution on a transverse plane in the far-field of azimuthal Walsh filters is determined by Eq. (2.16). The real and imaginary parts of the complex amplitude distribution are determined by the Eqs. (2.18) and (2.19) which are the series of Bessel functions of odd and even arguments of reduced transverse distance p as given in Eq. (2.9). The inherent phase asymmetry imposed by azimuthal Walsh filters placed on exit pupil plane leads to asymmetrical amplitude point spread function in the far-field plane.

The variation of real part, $C(p, \zeta; \theta)$ of complex amplitude distribution $A(p, \zeta)$ has been plotted against reduced transverse distance p for azimuths $\zeta = 0°$, $45°$, $90°$, $135°$, $180°$, $225°$, $270°$, $315°$ and $360°$ in the far-field plane (Fig. 2.4). It is seen that $C(p, \zeta; \theta)$ is either independent of azimuthal variation, ζ or maintains constant values for different values of ζ defined in the far-field plane.

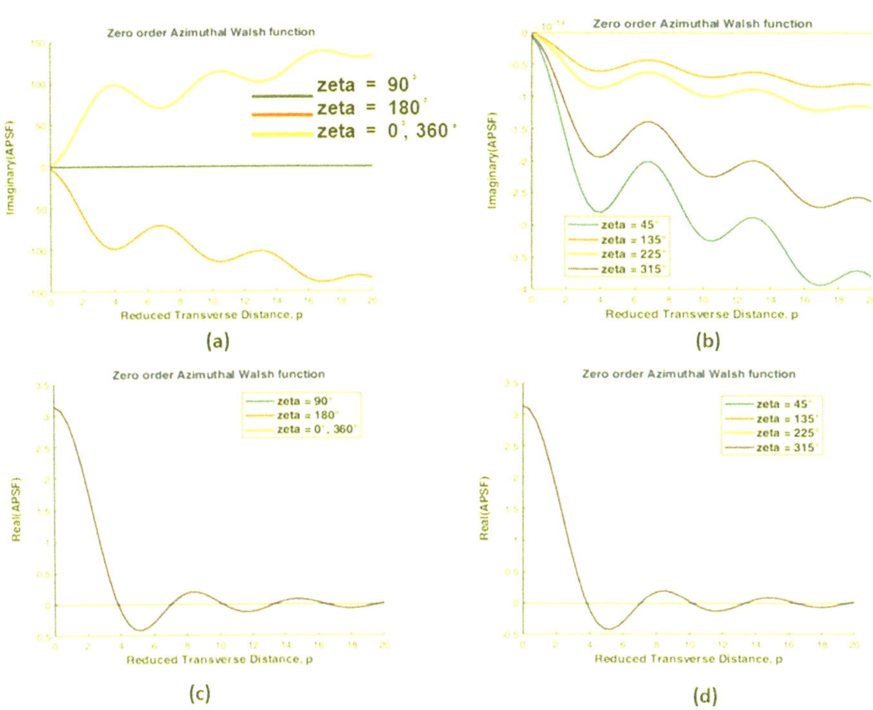

Fig. 2.4 Variation of imaginary and real part of complex amplitude distribution with azimuthal angle ζ at far-field plane for Zero order azimuthal Walsh filter at the exit pupil: **a** and **c** for $\zeta = 0°$, $90°$, $180°$ and $360°$; **b** and **d** for $\zeta = 45°$, $135°$, $225°$ and $315°$

The variation of imaginary part $S(p, \zeta; \theta)$ of complex amplitude distribution, $A(p, \zeta)$ has been plotted against reduced diffraction variable p for different values of azimuths, ζ namely, $\zeta = 0°$, 45°, 90°, 135°, 180°, 225°, 270° 315° and 360° in the far-field plane. The imaginary part of complex amplitude distribution shows pronounced azimuthal dependence. The variation of real and imaginary parts of complex amplitude distributions as well as variation of IPSF with azimuthal angle for zero, first, second and third order azimuthal Walsh filters on the exit pupil plane are discussed in the following sub-sections.

2.4.1 Case 1: Zero Order Azimuthal Walsh Filter

Figure 2.4 shows that the imaginary part of complex amplitude point spread function is widely varying with change in far-field angle ζ. Contribution of imaginary part in far-field amplitude distribution for $\zeta = 0°$ and 360° is just equal and opposite to that for $\zeta = 180°$ and hence cancels each other. Contributions of imaginary part in far-field amplitude distribution for $\zeta = 45°$, 135°, 225° and 315° are very much negligible, of the order of 10^{-14}. The contribution of real part of complex amplitude distribution in the far-field plane for zero order azimuthal Walsh filter on exit pupil is positive and independent of change in values of far-field angle ζ as illustrated in Fig. 2.4.

Intensity distribution at the far-field plane is plotted against reduced transverse distance, p for zero order azimuthal Walsh filter on the exit pupil plane as shown in Fig. 2.5. The intensity distribution is an Airy pattern which closely resembles with the real part of the complex amplitude distribution and is completely independent of azimuthal variation. The centre of the intensity distribution is maxima and the pattern is symmetrical about the origin.

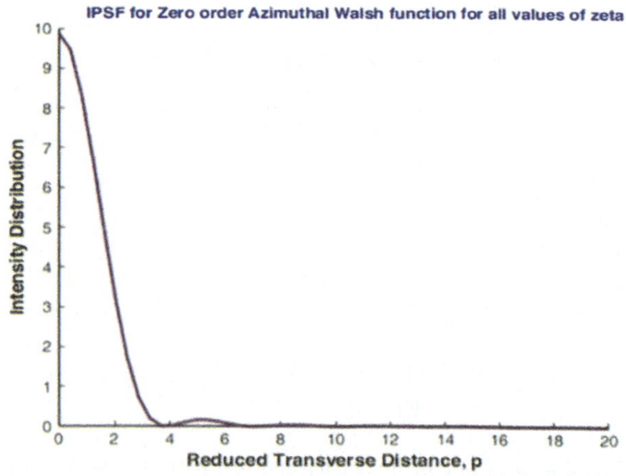

Fig. 2.5 Far-field intensity distribution for zero order azimuthal walsh filter on exit pupil shows symmetry against variation of zeta (ζ)

2.4.2 Case 2: First Order Azimuthal Walsh Filter

The variations of real and imaginary parts of complex amplitude point spread functions with azimuthal angle, ζ in the far-field plane are shown in Fig. 2.6 for first order azimuthal Walsh filter placed on the exit pupil plane. Contribution of imaginary part in the complex far-field amplitude distribution for $\zeta = 90^\circ$ is appreciable. Contributions of imaginary part in the far-field amplitude distribution for other values of ζ are considerable. On the other hand, contributions of real part in the complex far-field amplitude distribution for change in far-field angle ζ, is very much negligible, of the order of 10^{-15} and maintains constant values for any particular value of far-field angle ζ as illustrated in Fig. 2.6. Hence it can be inferred that contribution of real part is becoming negligible for forming the far-field amplitude distribution pattern for first order azimuthal Walsh filters placed at the exit pupil plane. The contribution

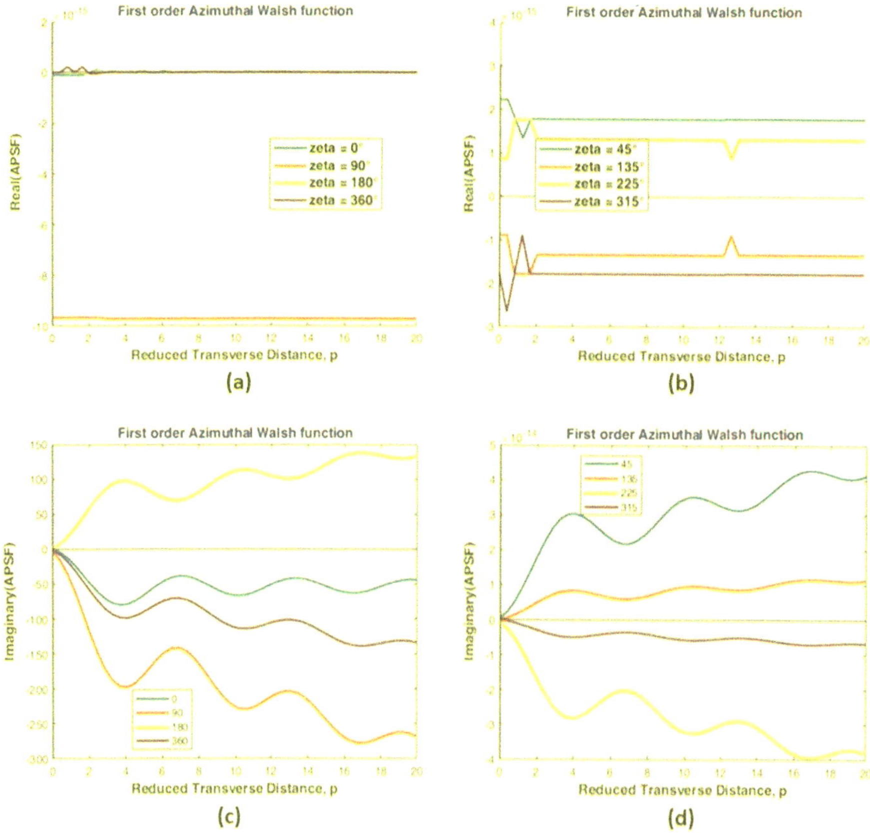

Fig. 2.6 Variation of imaginary and real part of complex amplitude distribution with azimuthal angle ζ at far-field plane for first order azimuthal walsh filter at the exit pupil: **a** and **c** for $\zeta = 0^\circ$, 90°, 180° and 360°; **b** and **d** for $\zeta = 45^\circ$, 135°, 225° and 315°

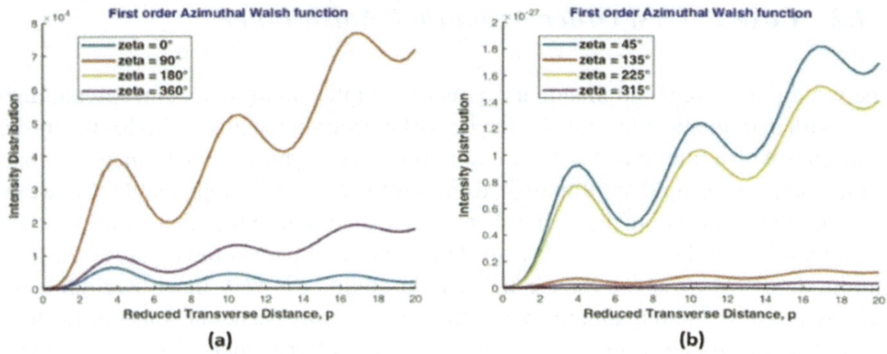

Fig. 2.7 Far-field intensity distribution for first order azimuthal walsh filter **a** for $\zeta = 0°, 90°, 180°$ and 360°; **b** for $\zeta = 45°, 135°, 225°$ and 315°

of imaginary part is predominant in the complex amplitude distribution which shows the azimuth dependence imposed by azimuthal Walsh filter of order one.

Intensity distributions at the far-field plane are plotted against reduced transverse distance, p for First order azimuthal Walsh is shown in Fig. 2.7. The intensity distribution is changing widely with change in the values of ζ in the far-field plane which is due to the phase asymmetry imposed by the Walsh filter. The resultant intensity distribution at the focal plane for all values of ζ within $(0, 2\pi)$ will be asymmetrical compared to the symmetrical Airy pattern produced by zero order azimuthal Walsh filter. The centre of the intensity distribution for all values of ζ show minima.

2.4.3 *Case 3: Second Order Azimuthal Walsh Filter*

The variations of real and imaginary parts of complex amplitude point spread functions with azimuthal angle, ζ in the far-field plane for Second order azimuthal Walsh filters placed on the exit pupil plane are shown in Fig. 2.8. Contributions of imaginary parts in far-field amplitude distribution for $\zeta = 0°$, 180° and 360° are found appreciable. On the other hand, the contributions of real part of far-field complex amplitude distributions are shown to be very minimal, of the order of 10^{-13}–10^{-15} or sometimes constant for any particular value of ζ as illustrated in Fig. 2.8. The corresponding intensity distributions for are plotted against reduced transverse distance and shown in Fig. 2.9.

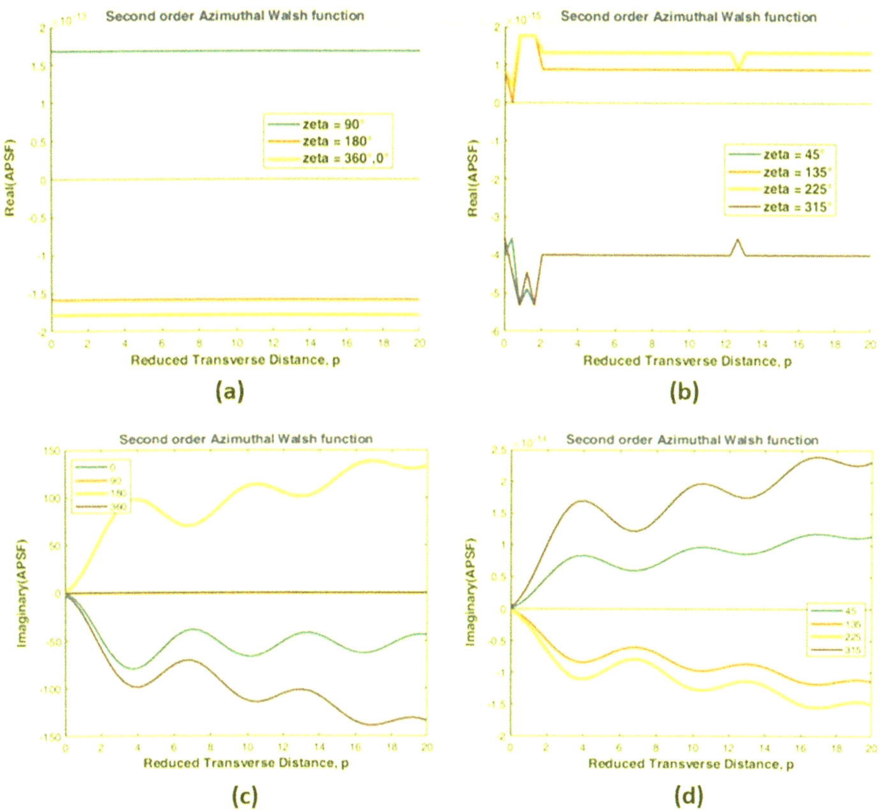

Fig. 2.8 Variation of imaginary and real part of complex amplitude distribution with azimuthal angle ζ at far-field plane for Second order azimuthal walsh filter at the exit pupil: **a** and **c** for ζ = 0°, 90°, 180° and 360°; **b** and **d** for ζ = 45°, 135°, 225° and 315°

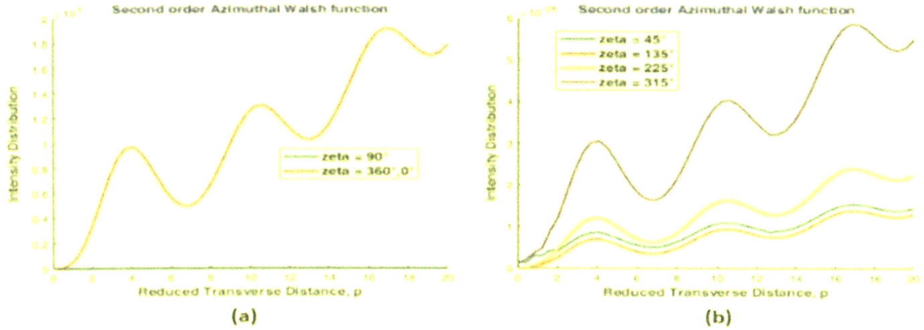

Fig. 2.9 Far-field intensity distribution for second order azimuthal walsh filter on exit pupil: **a** for ζ = 0°, 90° and 360°; **b** For ζ = 45°, 135°, 225° and 315°

2.4.4 Case 4: Third Order Azimuthal Walsh Filter

For third order azimuthal Walsh filter placed on the exit pupil plane of a rotationally symmetric imaging system, imaginary part of far-field amplitude distribution for different values of ζ ranging from 0°, 45°, 90°, 135°, 180°, 225°, 315° and 360° are plotted against reduced transverse distance, p in Fig. 2.10. Variation of real part of far-field complex amplitude distribution for change in far-field angle ζ is plotted against p and shown in Fig. 2.10. As apparent from Fig. 2.10 the variations of real part maintain constant values for different values ζ. The contribution of imaginary part is predominant in the intensity point spread function as illustrated in Fig. 2.11.

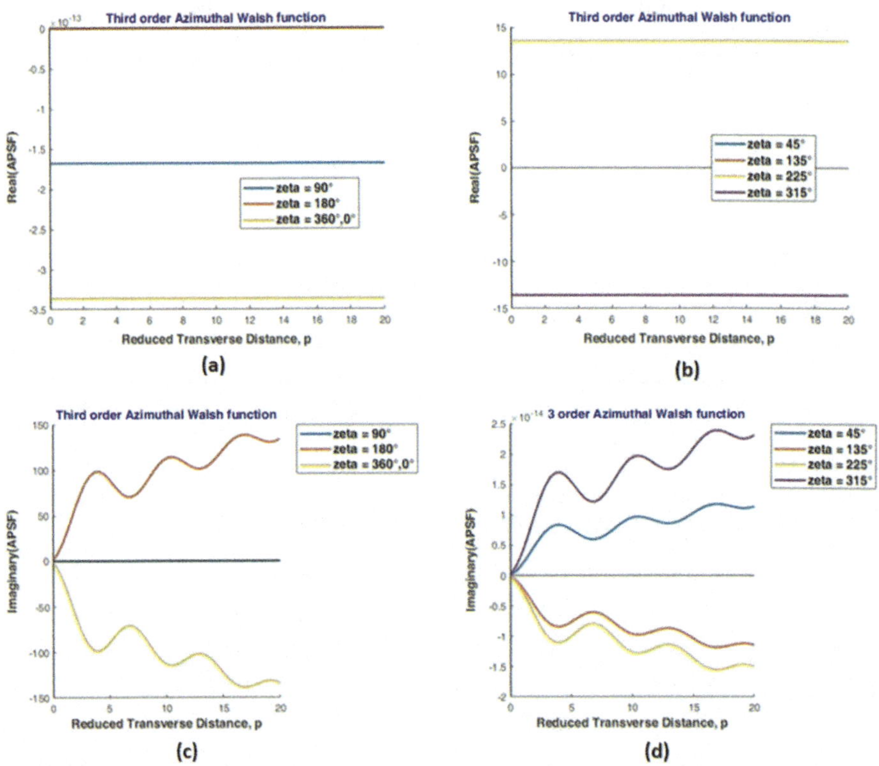

Fig. 2.10 Variation of imaginary and real part of complex amplitude distribution with azimuthal angle ζ at far-field plane for second order azimuthal walsh filter at the exit pupil: **a** and **c** for $\zeta =$ 0°, 90°, 180° and 360°; **b** and **d** for $\zeta = 45°$, 135°, 225° and 315°

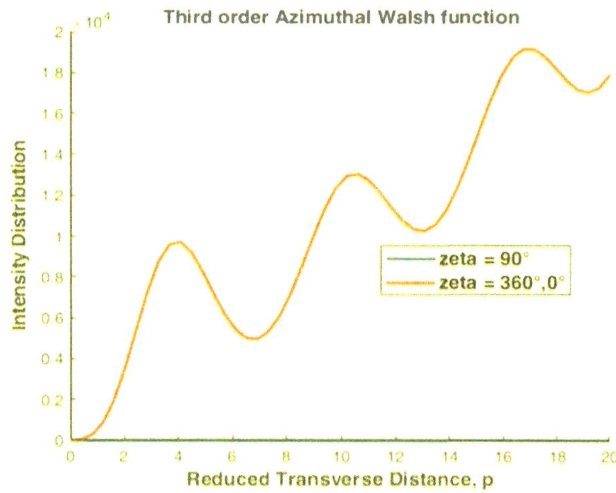

Fig. 2.11 Far-field intensity distribution for third order azimuthal Walsh filter on exit pupil for $\zeta = 0°$, $90°$ and $360°$

2.5 2D Intensity Distribution on the Far-Field Plane

The azimuthal Walsh filter is placed on the exit pupil of a rotationally symmetric imaging system (Fig. 2.2) and the normalized 2D intensity distribution on a transverse plane in the far-field of azimuthal Walsh filters is determined by Eq. (2.22).

A point on the transverse image plane in the far-field is represented by the Cartesian co-ordinate (ξ, η) with O' as the origin of the axes as in Fig. 2.3. The normalized transverse intensity $I_N(p, \zeta)$ is represented by Eq. (2.22). where (p, ζ) are the reduced diffraction variables defined as the reduced transverse distance, p as represented in Eq. (3.8) and ζ is the azimuthal variation. Normalized transverse intensity $I_N(p, \zeta)$ against variation of reduced transverse distance, p at different azimuthal variation, ζ has been plotted for all orders of azimuthal Walsh filters from 0 to 63. The 2D normalized intensity variations along with contour plots are shown for each individual order of azimuthal Walsh filters. Except for zero order, the normalized intensity distribution against reduced transverse distance show asymmetrical variation along different azimuths.

For zero order azimuthal Walsh filter at the exit pupil, the transverse intensity distribution on the far-field plane is the Airy pattern which is symmetric and showing no azimuthal dependence. The normalized intensity distribution with the central maxima at the origin of the far-field plane with magnitude one is illustrated in Fig. 2.12.

The intensity distributions on the far-field plane corresponding to azimuthal Walsh filters $W_\nu(\theta)$, for orders $\nu = 0, 1, 2, 3, \ldots, 63$ are shown in Table 2.1 in the next page. A contour plot of the intensity distribution is also shown in each case to underscore the nature of intensity variations.

The Far-field intensity distributions for azimuthal Walsh filters for orders 0–63 are evaluated analytically and presented in Table 2.1. The intensity distributions along with the corresponding contour plot on the far-field plane are shown for each order ν

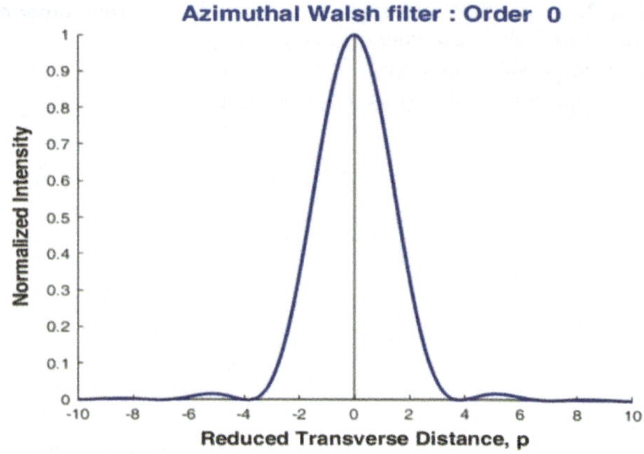

Fig. 2.12 The normalized intensity distribution at far-field plane due to zero order azimuthal Walsh filter on exit pupil, the Airy pattern

of azimuthal Walsh filter $W_{\nu}(\theta)$. The contour plots corresponding to intensity distribution has been made coloured to highlight the position of the peak or maximum value compared to other values of intensity. The yellow colour represents the location of peak values or maxima of the intensity distribution pattern obtained for any azimuthal Walsh filter $W_{\nu}(\theta)$ of order ν. It is apparent from 2D IPSF that the central dark region of the intensity distribution spreads out as order ν increases.

A 3D view of the IPSF can be generated from the 2D pattern, as presented for zero order azimuthal Walsh filter, $W_0(\theta)$ in Table 2.2. The intensity distribution is showing central maxima with alternate dark and bright concentric rings of diminishing amplitude arranged in a symmetric fashion about the centre and is independent of azimuthal variation along the far-field plane. The central yellow patch shows the location of peak value or the maxima of the intensity distribution.

A 3D view of the IPSF for first and second order azimuthal Walsh filters, $W_1(\theta)$ and $W_2(\theta)$ is presented in Table 2.3. The intensity distribution is asymmetric, and it is dependent upon the azimuthal variation along the far-field plane with central minima. The yellow patches show the location of peak values or the maxima of the intensity distribution. It is seen that the peak values are located on both sides of the origin.

2.6 Experimental Verification

Figure 2.13 shows the experimental arrangement for addressing reflecting SLMs to codify with azimuthal Walsh filters of different orders and corresponding far-field diffraction patterns are generated and recorded.

Table 2.4 shows the experimental results of far-field intensity distributions for some specific orders of azimuthal Walsh filters and compared with corresponding patterns generated analytically. The results show perfectly matched with each other.

Table 2.1 Azimuthal Walsh filters $W_{\nu}(\theta)$, for orders $\nu = 0, 1, 2, \ldots, 63$ and their transverse intensity distributions on far-field plane alongwith intensity contour diagrams

Order	Azimuthal Walsh filters	2D IPSF	Contour IPSF	Scale
0		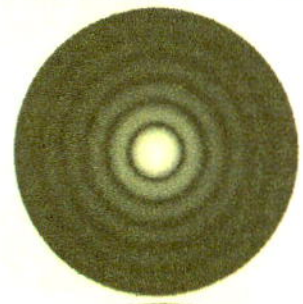	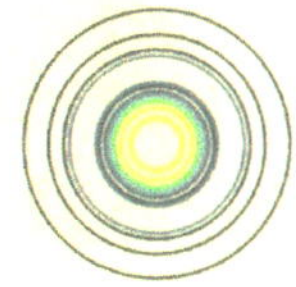	
1		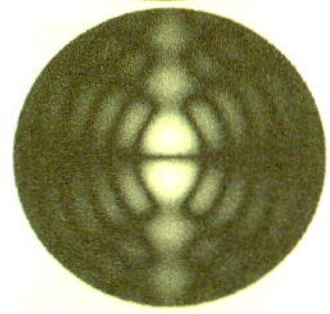	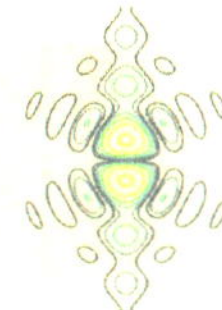	
2	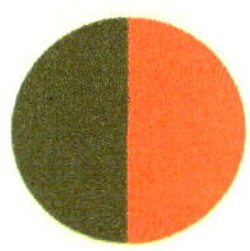	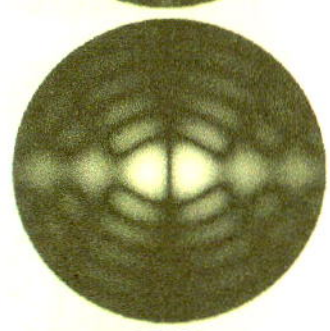	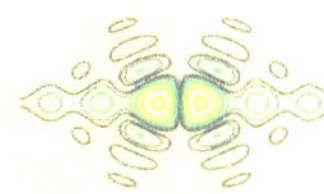	
3		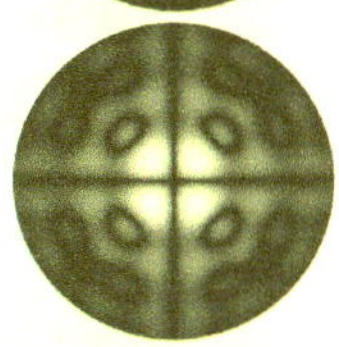		
4		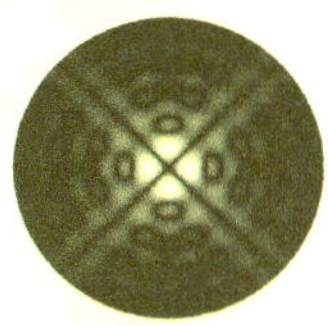	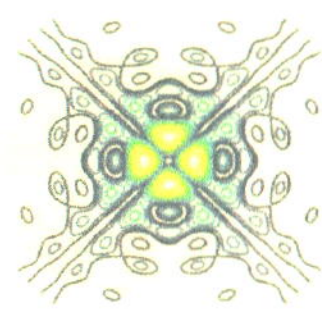	
5				
6		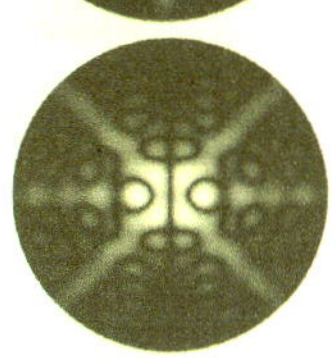	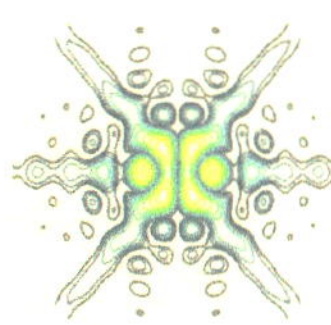	

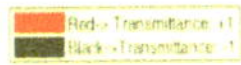

(continued)

Table 2.1 (continued)

Order	Azimuthal Walsh filters	2D IPSF	Contour IPSF	Scale
7		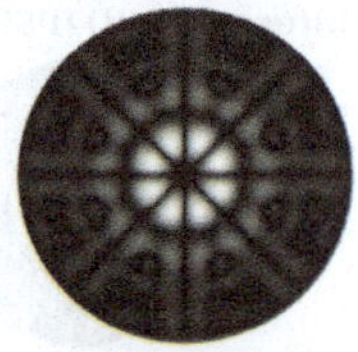		
8		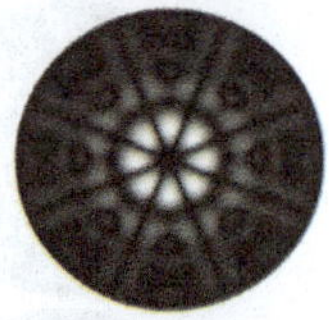	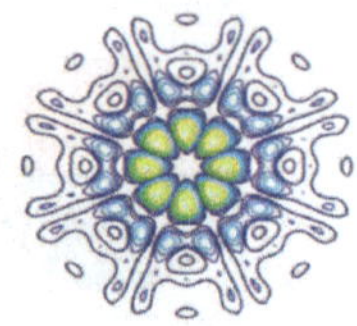	
9		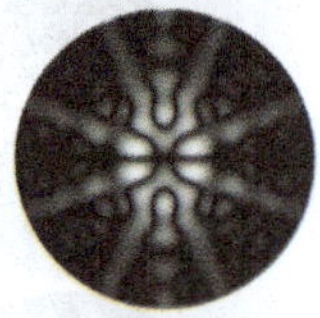	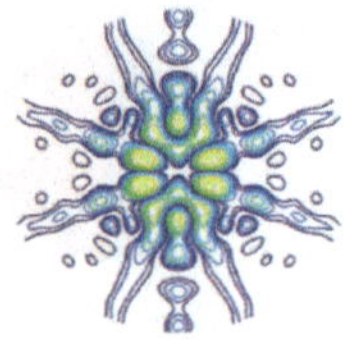	
10			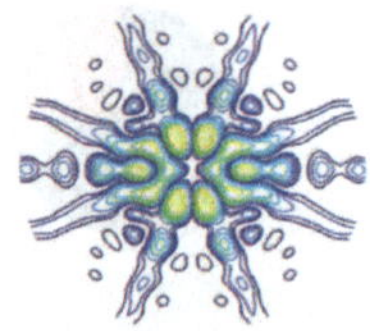	
11			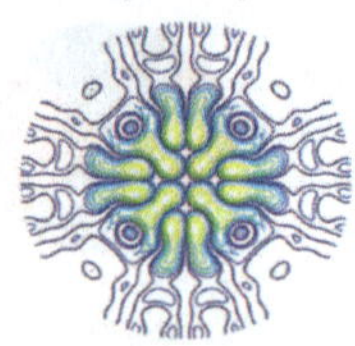	
12				
13			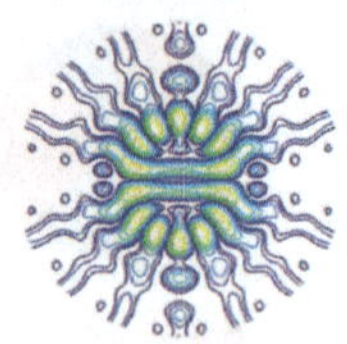	

(continued)

Table 2.1 (continued)

Order	Azimuthal Walsh filters	2D IPSF	Contour IPSF	Scale
14		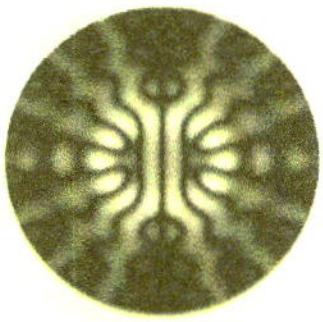	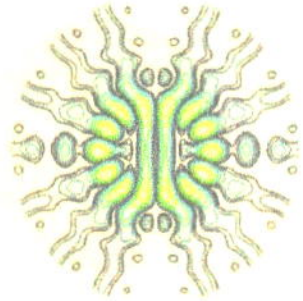	
15		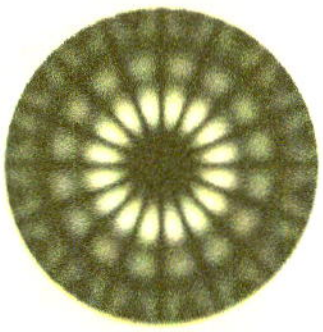		
16		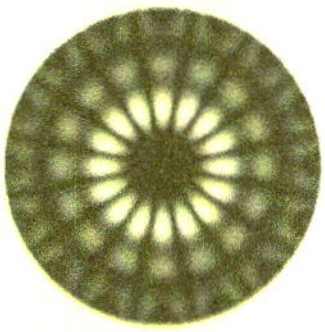		
17		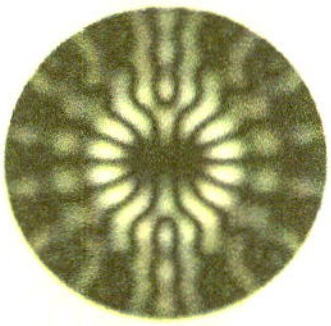	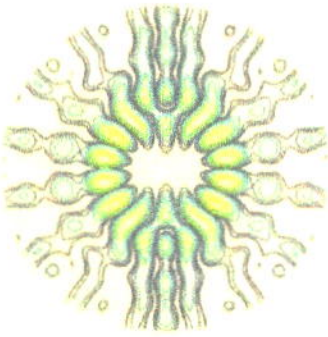	
18			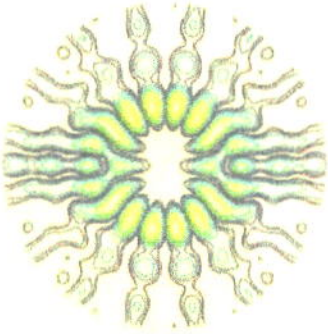	
19		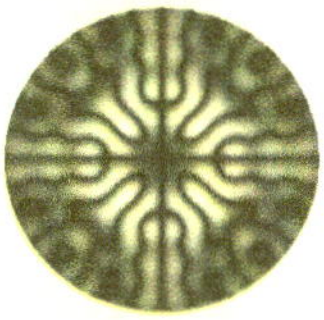	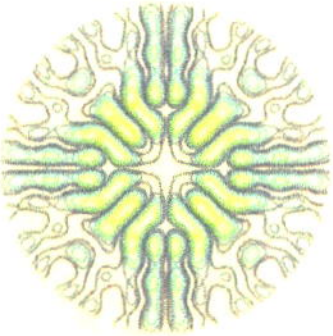	
20		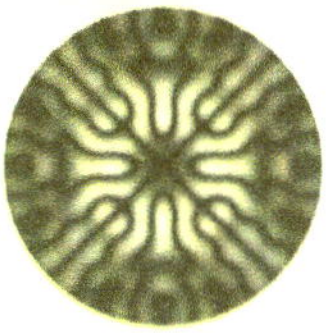	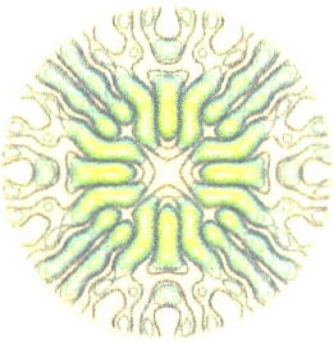	

(continued)

Table 2.1 (continued)

Order	Azimuthal Walsh filters / 2D IPSF / Contour IPSF	Scale
21	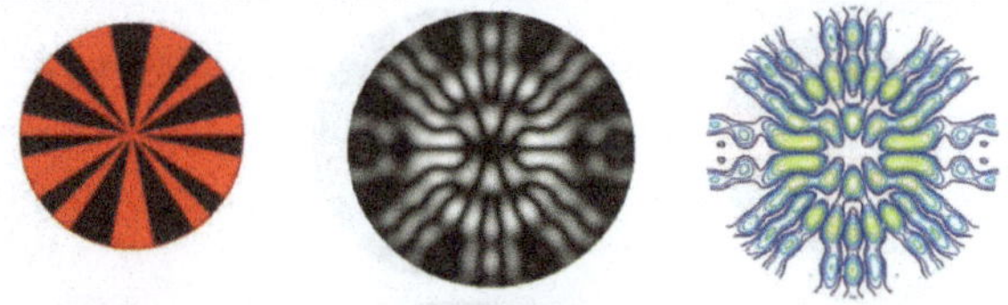	
22	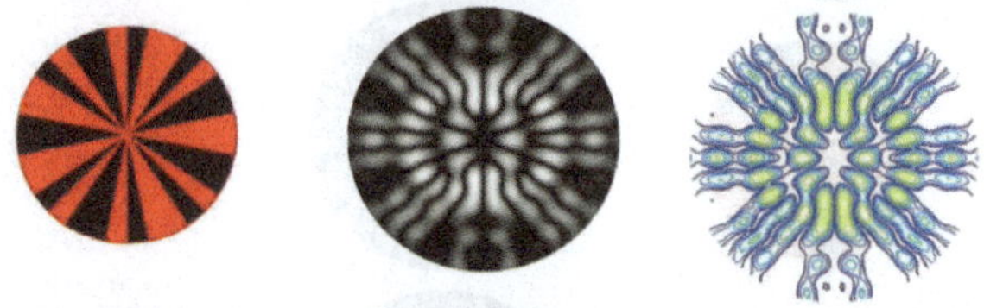	
23	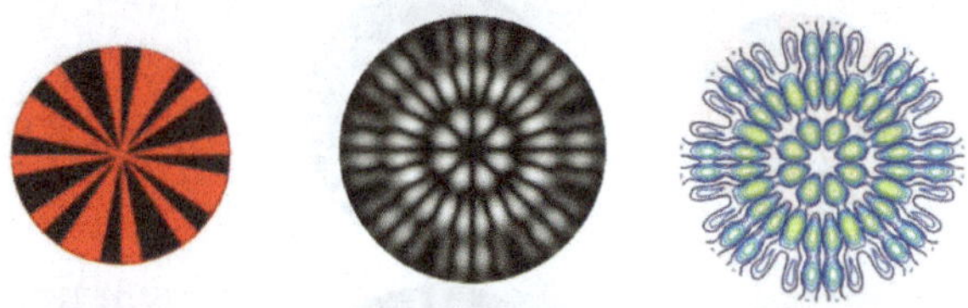	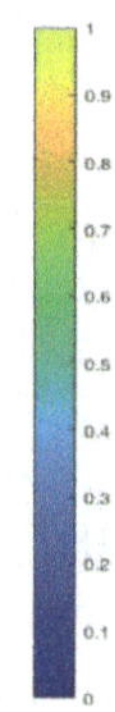
24	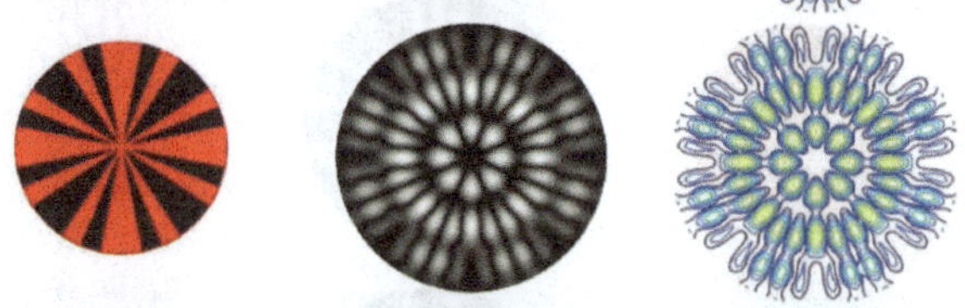	
25	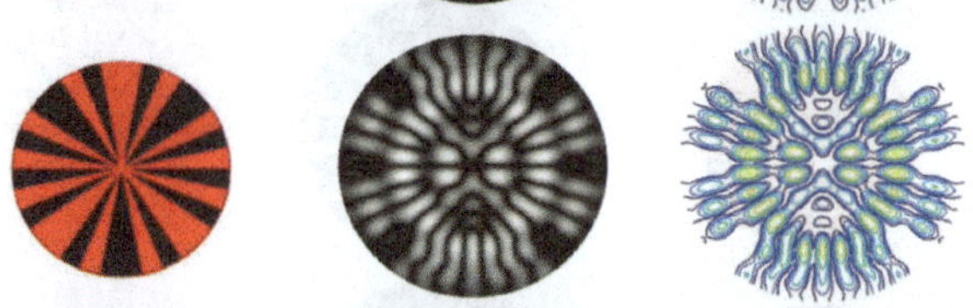	
26	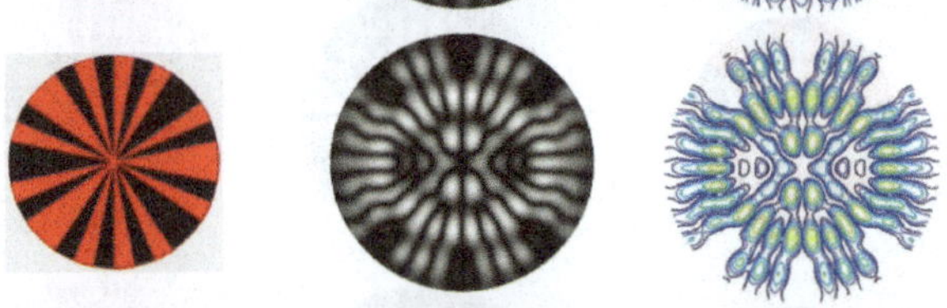	
27	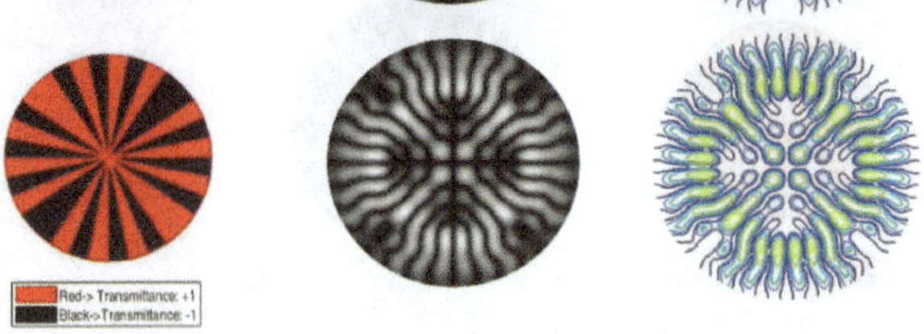	

(continued)

Table 2.1 (continued)

Order	Azimuthal Walsh filters	2D IPSF	Contour IPSF	Scale
28		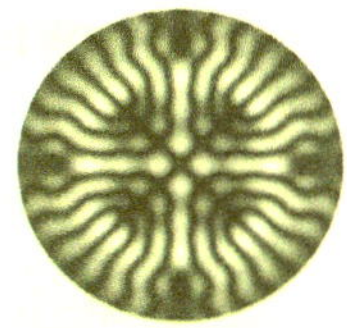	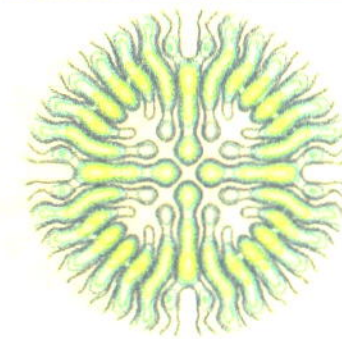	
29		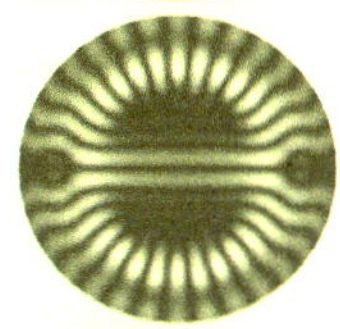	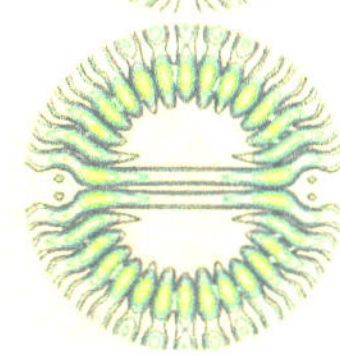	
30		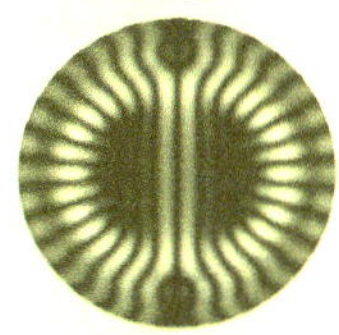	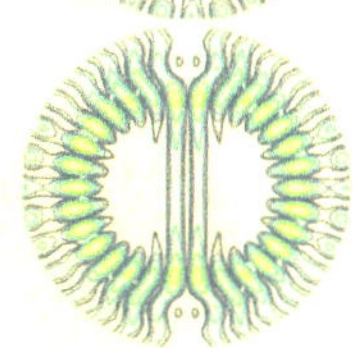	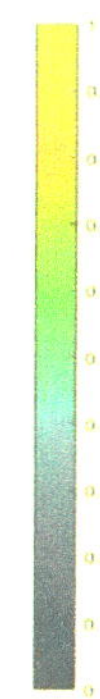
31		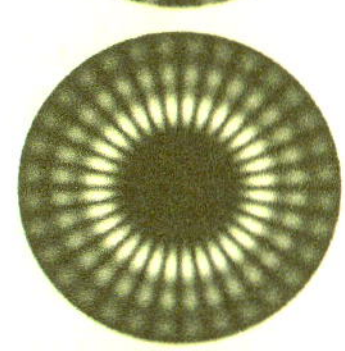	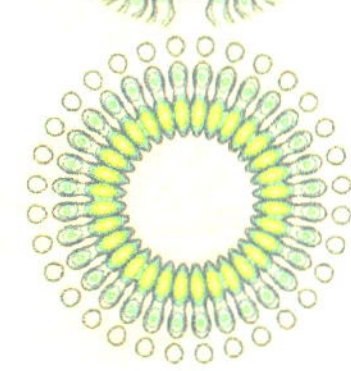	
32		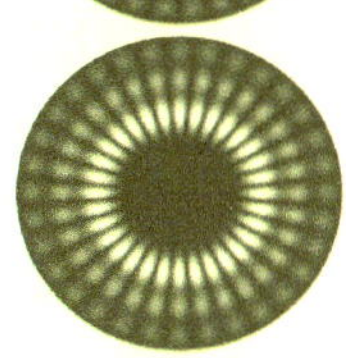	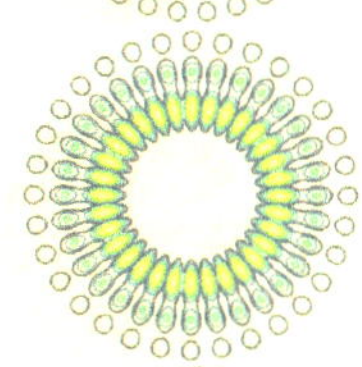	
33		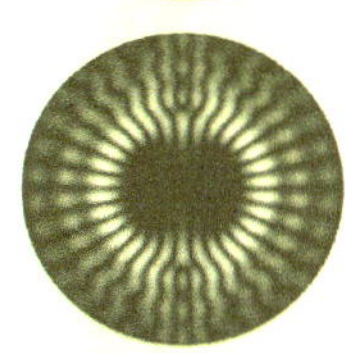	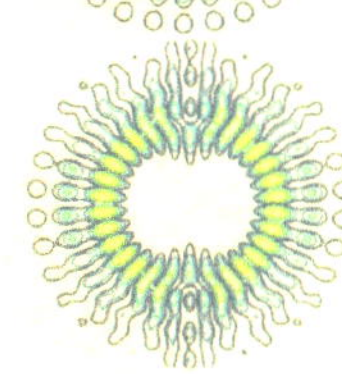	
34		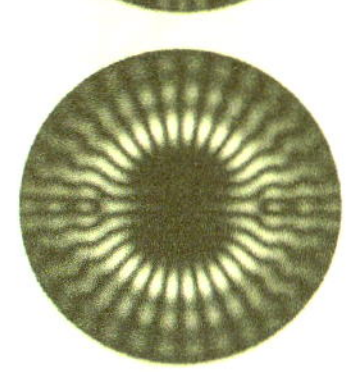	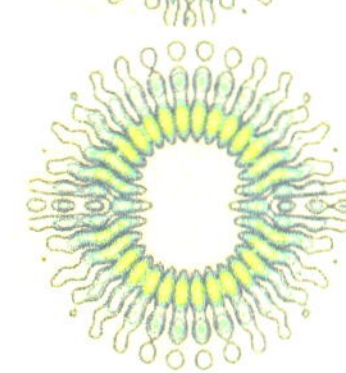	

(continued)

Table 2.1 (continued)

Order	Azimuthal Walsh filters	2D IPSF	Contour IPSF	Scale
35			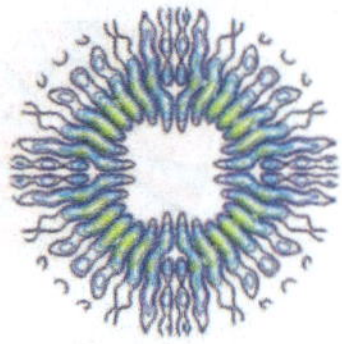	
36		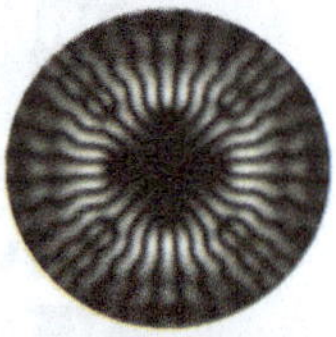	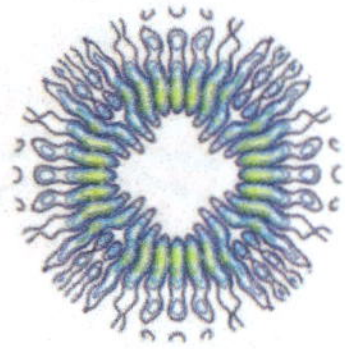	
37		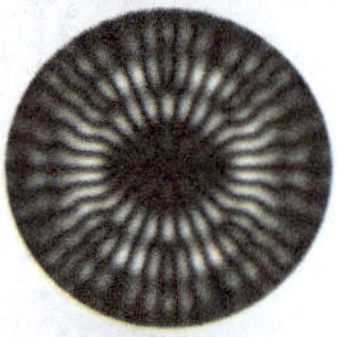	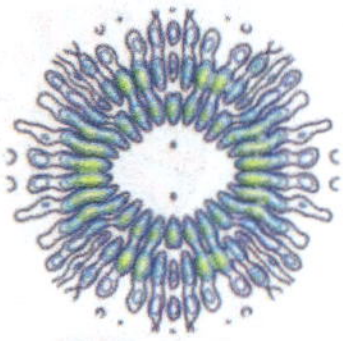	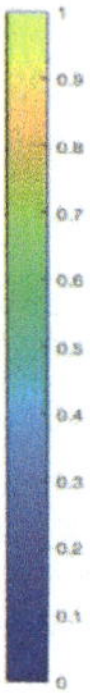
38			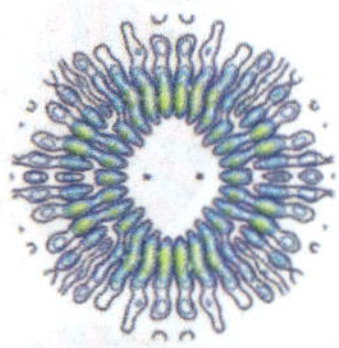	
39		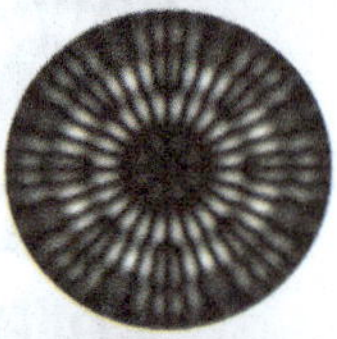	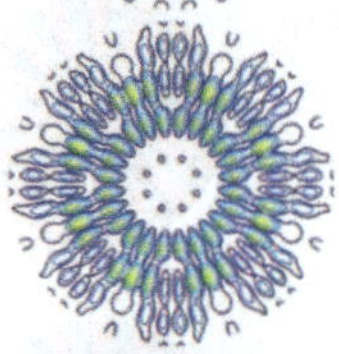	
40	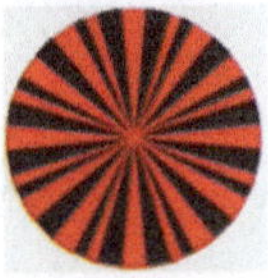	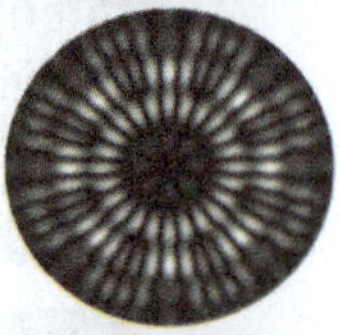	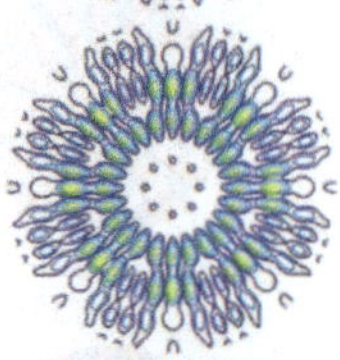	
41			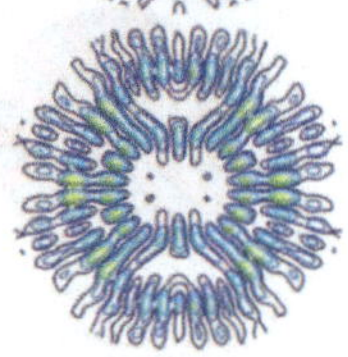	

(continued)

Table 2.1 (continued)

Order	Azimuthal Walsh filters	2D IPSF	Contour IPSF	Scale
42		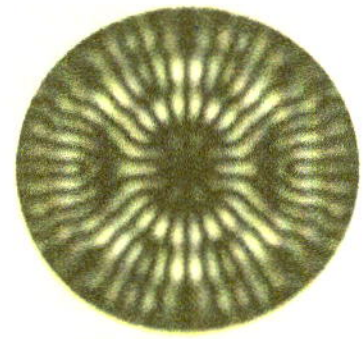	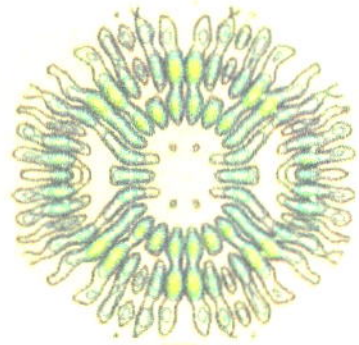	
43		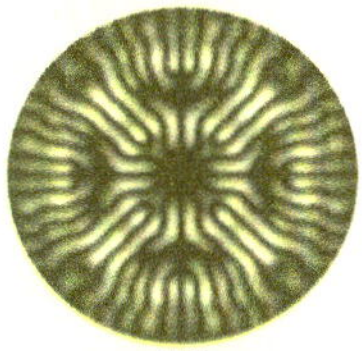	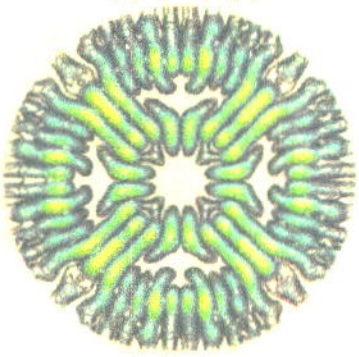	
44		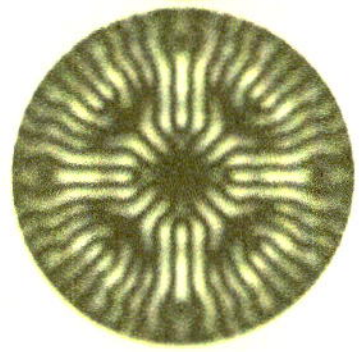	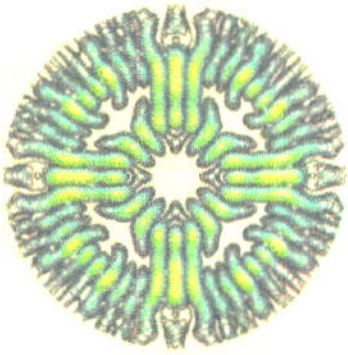	
45		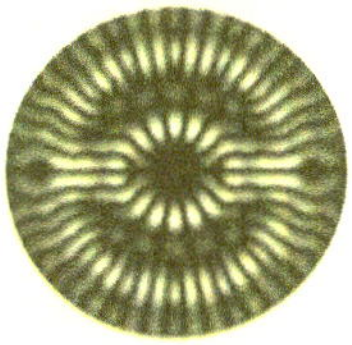	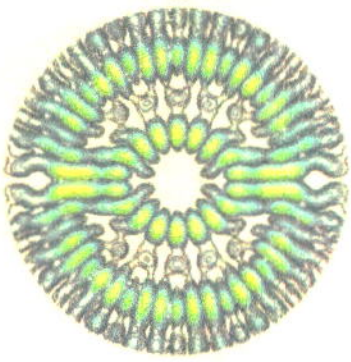	
46		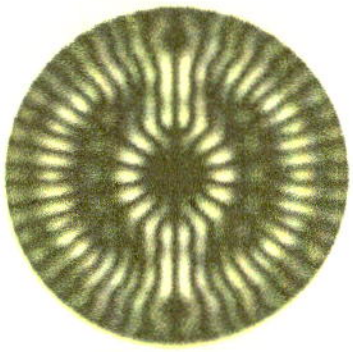	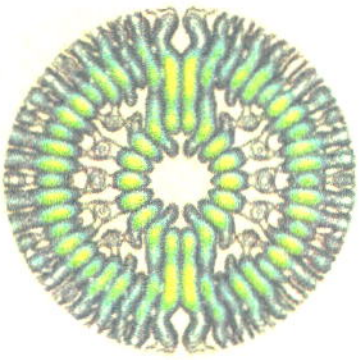	
47	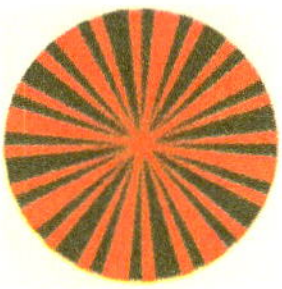	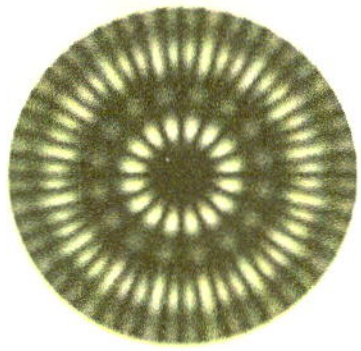		
48	 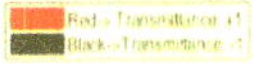	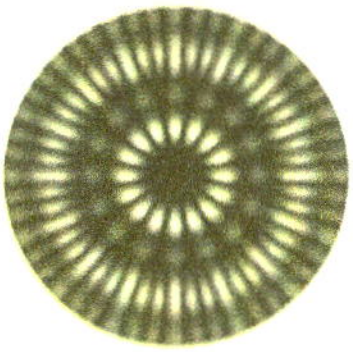		

(continued)

Table 2.1 (continued)

Order	Azimuthal Walsh filters	2D IPSF	Contour IPSF	Scale
49		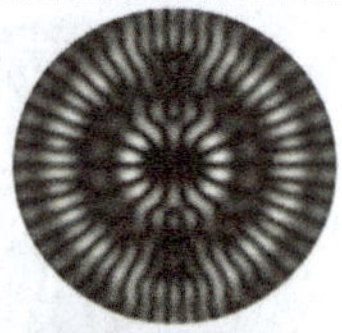	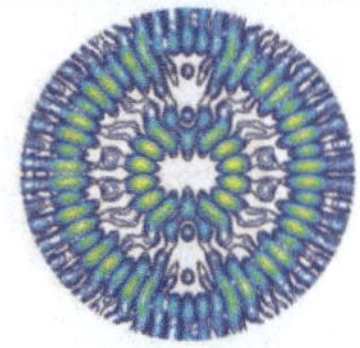	
50		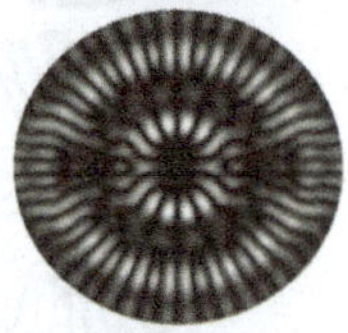		
51		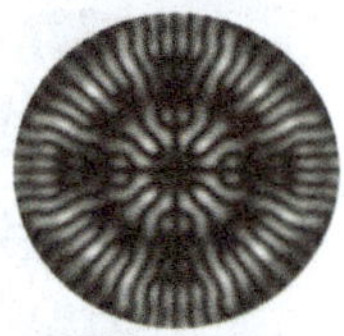	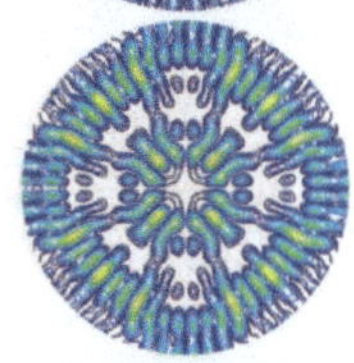	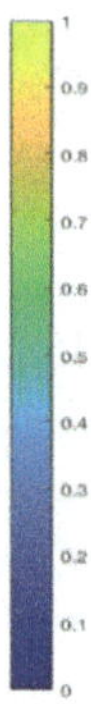
52		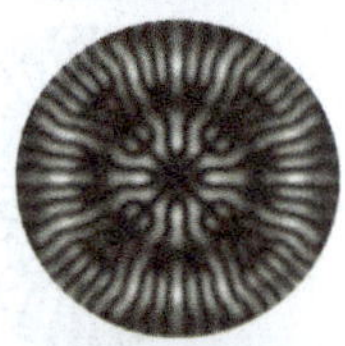	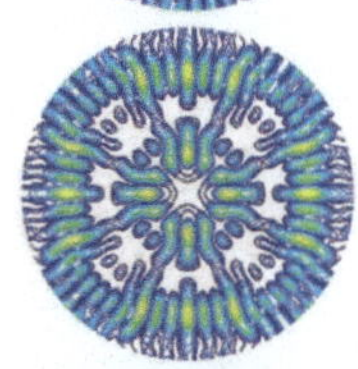	
53		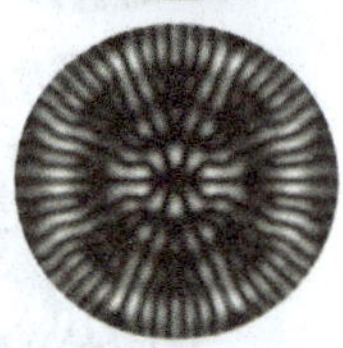	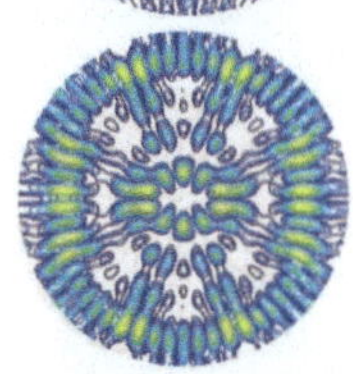	
54		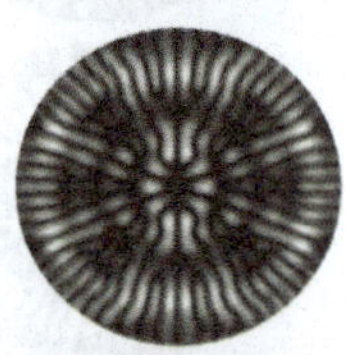	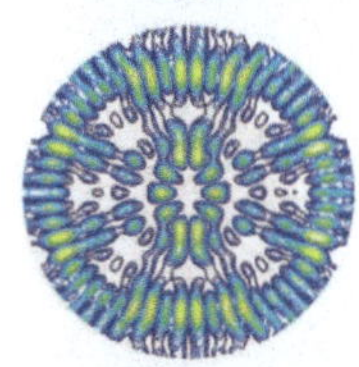	
55	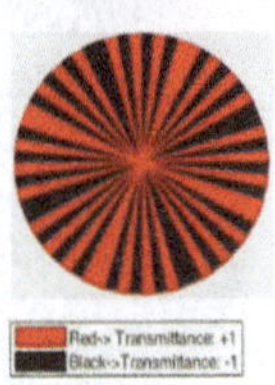	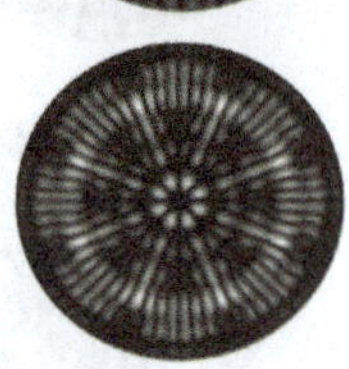		

(continued)

Table 2.1 (continued)

Order	Azimuthal Walsh filters	2D IPSF	Contour IPSF	Scale
56	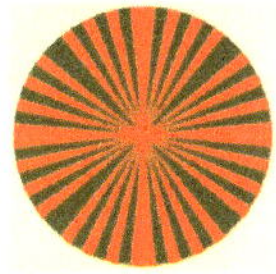	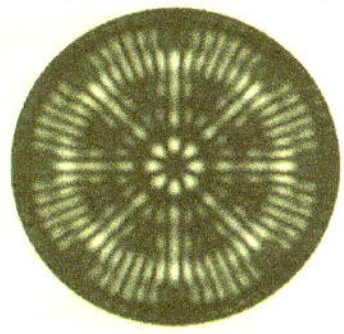	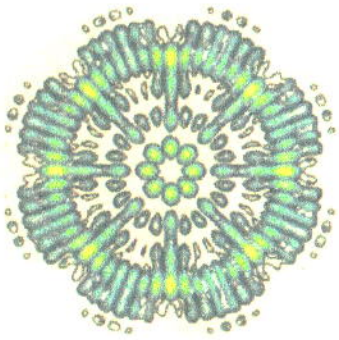	
57		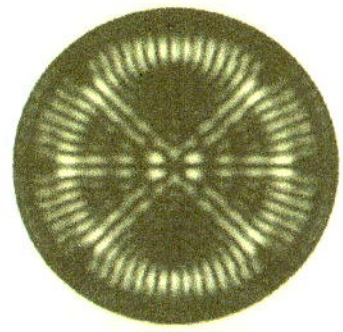	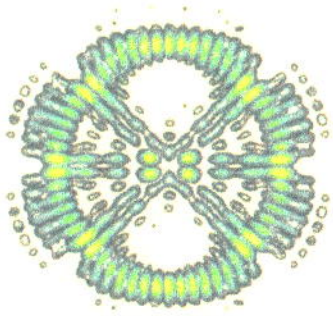	
58		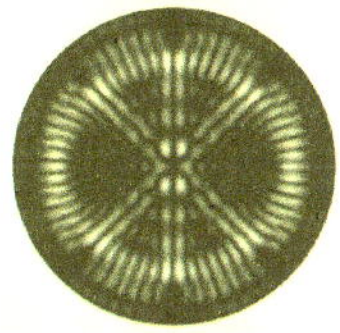	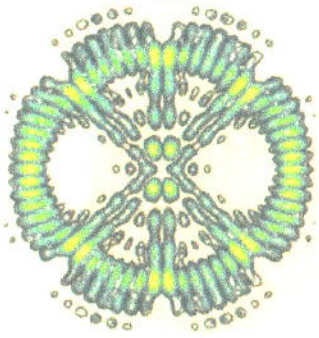	
59		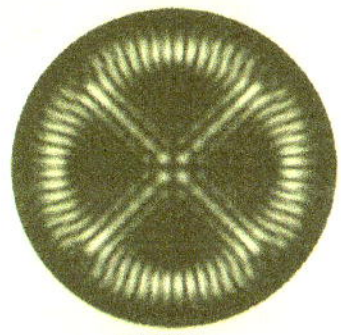	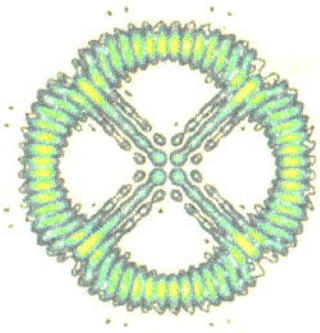	
60		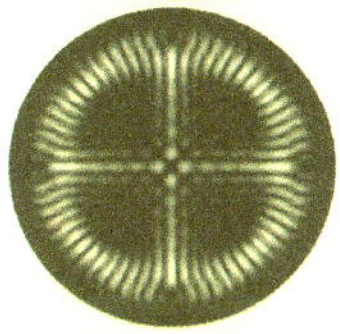		
61			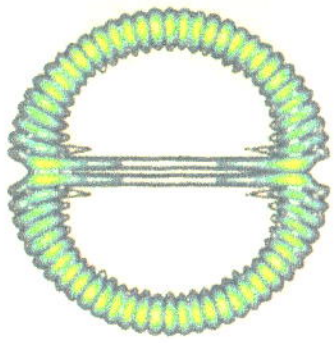	
62	 	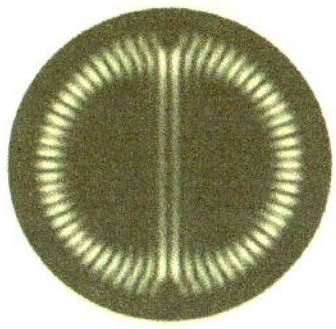	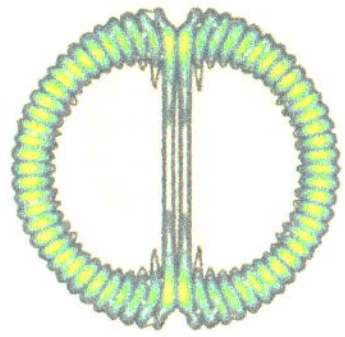	

(continued)

Table 2.1 (continued)

Order	Azimuthal Walsh filters	2D IPSF	Contour IPSF	Scale
63	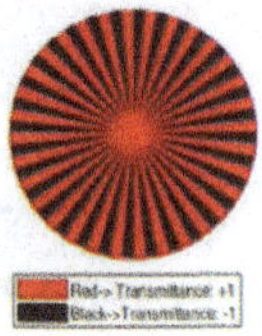		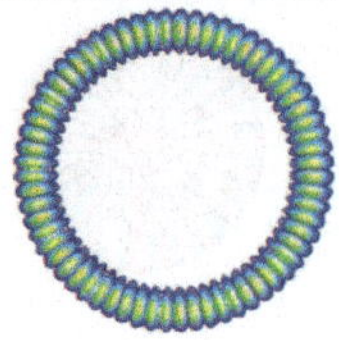	

Table 2.2 Azimuthal Walsh filter $W_\nu(\theta)$, for order $\nu = 0$ and its 3D view of 2D transverse intensity distribution at far-field plane is generated analytically

Order	Azimuthal Walsh filters	Contour IPSF	3D IPSF generated	Scale
0	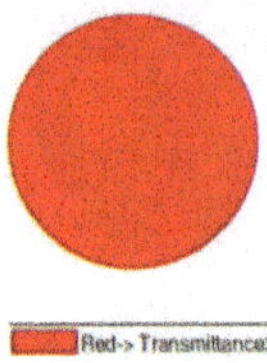	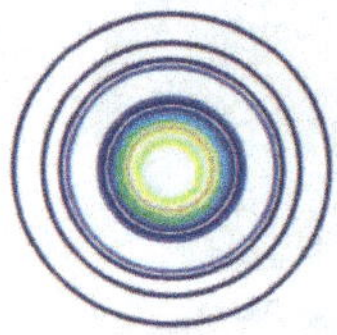	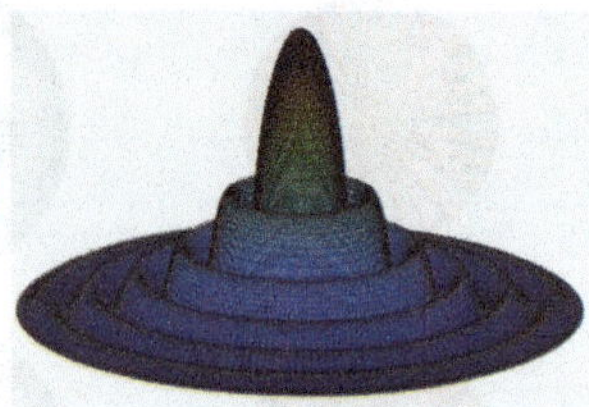	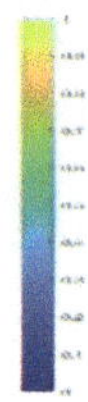

Table 2.3 Azimuthal Walsh filter $W_\nu(\theta)$, for orders $\nu = 1\&2$ and snapshots of the 3D views of transverse intensity distributions at far-field plane generated analytically

Order	Azimuthal Walsh filters	Contour IPSF	3D IPSF generated	Scale
1				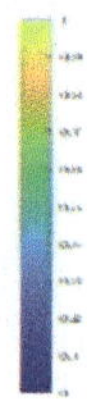
2	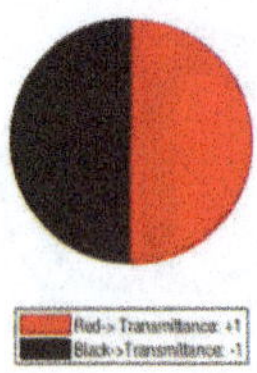	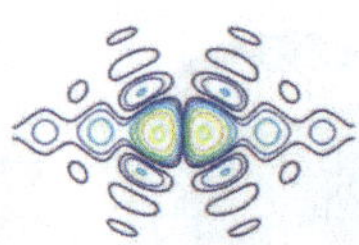		

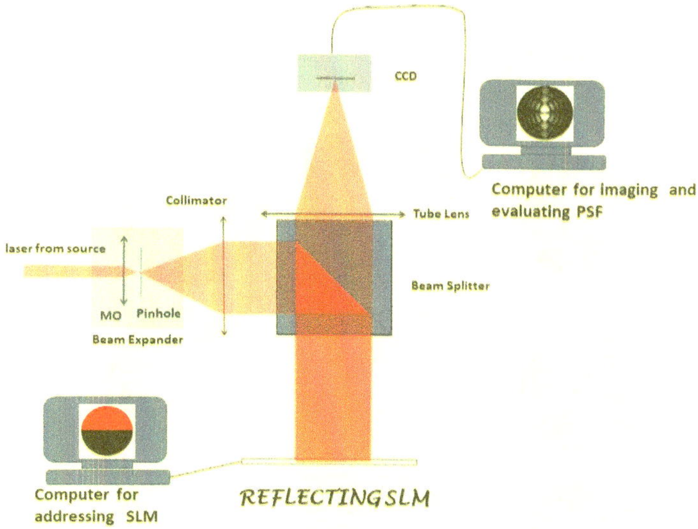

Fig. 2.13 Schematic diagram of the experimental arrangement showing reflecting type spatial light modulator (SLM) to generate azimuthal Walsh filters and their corresponding far-field diffraction patterns are recorded

Table 2.4 2D IPSFs generated analytically at the far-field plane are matching with the experimental results for random ordered azimuthal Walsh filters

Order	Azimuthal Walsh filters	2D IPSF		Scale
		Experimentally Obtained	Analytically Calculated	
0			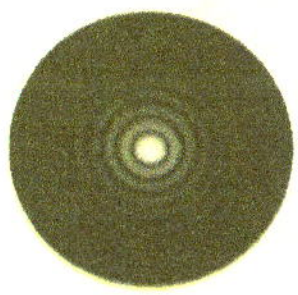	
1		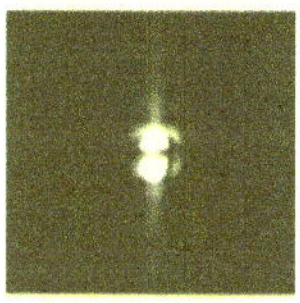		
3			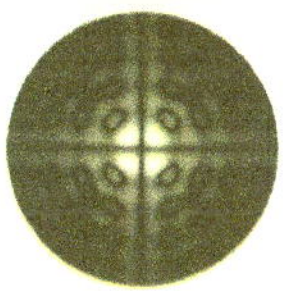	

(continued)

Table 2.4 (continued)

Order	Azimuthal Walsh filters	2D IPSF Experimentally Obtained	Analytically Calculated	Scale
5		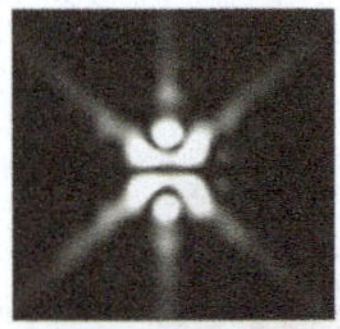		
7			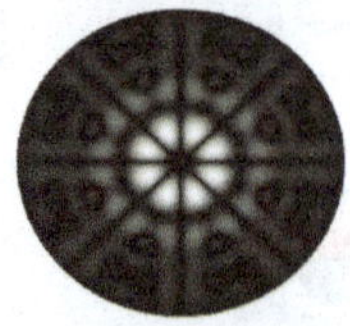	
10			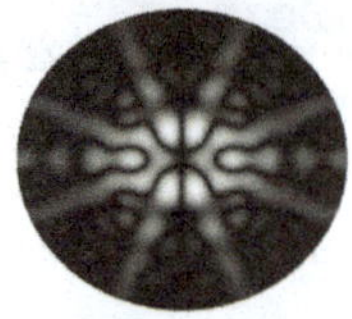	
11				
14				

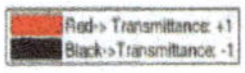

References

1. G. Toraldo di Francia, La Diffrazione Delle Luce, Edizioni Scientifiche Einaudi (1968)
2. G. Toraldo di Francia, Super-gain antennas and optical resolving power. Nuovo. Cimento. Supple. **9**(3), 426–438 (1952)
3. G. Toraldo di Francia, Nuovo pupille superresolventi. Atti. Fond. Giorgio. Ronchi. **7**, 366–372 (1952)
4. L.N. Hazra, Walsh filters in tailoring of resolution in microscopic imaging. Micron **38**(2), 129–135 (2007)
5. L.N. Hazra, A new class of optimum amplitude filters. Opt. Commun. **21**(2), 232–236 (1977)

6. L.N. Hazra, P.K. Purkait, M. De, Apodization of aberrated pupils. Can. J. Phys. **57**(9), 1340–1346 (1979)
7. P.K. Purkait, Application of walsh functions in problems of aberrated optical imagery. Ph.D. dissertation, University of Calcutta, India
8. L.N. Hazra, A. Guha, Far-field diffraction properties of radial Walsh filters. J. Opt. Soc. Am. A **3**, 843–846 (1986)
9. M. De, L.N. Hazra, Walsh functions in problems of optical imagery. Opt. Acta **24**, 221–234 (1977)
10. H. Lakshminarayan, P. Mukherjee, Optical masks using walsh functions, in ed. by R. Guenther, D. Steel *Encyclopedia of Modern Optics*, vol. 5, 2nd edn. (Elsevier, Oxford, 2018), pp. 14–29
11. P. Mukherjee, L.N. Hazra, Far-field diffraction properties of annular walsh filters. Adv. Opt. Technol. ID 360450 (2013)
12. O. Nakamura, K. Toyoda, Side lobe depression of the point-spread function in annular pupil optical systems. Appl. Opt. **30**(22), 3242–3245 (1991)
13. E.H. Linfoot, E. Wolf, Diffraction images in systems with annular aperture. Proc. Phys. Soc. B **66**, 145–149 (1953)
14. C.J.R. Sheppard, T. Wilson, Imaging properties of annular lenses. Appl. Opt. **18**(22), 3764–3769 (1979)
15. A. Boivin, *Théorie et calcul des figures de diffraction de revolution* (Gauthier-Villars, Paris, 1964)
16. W.T. Welford, Use of annular apertures to increase focal depth. J. Opt. Soc. Am. **50**, 749–753 (1960)
17. C.J.R. Sheppard, A. Choudhury, Annular pupils, radial polarization, and superresolution. Appl. Opt. **43**(22), 4322–4327 (2004)
18. M. Yun, M. Wang, L. Liu, Superresolution with annular binary phase filter in the 4Pi confocal system. J. Opt. A: Pure Appl. Opt. **7**(11), 640–644 (2005)
19. G.B. Airy, Trans. Cambridge Phil. Soc. **5**, 283 (1835)
20. M. Born, E. Wolf, *Principles of Optics* (Pergamon, Oxford, 1980)
21. A.I. Mahan, C.V. Bitterli, S.M. Cannon, Far-field diffraction patterns of Single and Multiple Apertures bounded by arcs and radii of concentric circles. J. Opt. Soc. Am. A **54**(6), 721–732 (1964)
22. R.A. Lessard, S.C. Som, Imaging properties of sector-shaped apertures. App. Opt., **II**(4), 811–817 (1972)
23. T. Vierke, Jahns Jürgen, Diffraction theory for azimuthally structured Fresnel zone plate. J. Opt. Soc. Am. A **31**(2), 363–372 (2014)
24. S. Mukhopadhyay, S. Sarkar, K. Bhattacharya, L.N. Hazra, Olarisation phase shifting interferometric technique for phase calibration of a reflective phase spatial light modulator. Opt. Eng. **52**(3), 035602-1–035602-6
25. I. Bhattacharya, A. Saha. L. Hazra, Asymmetrical PSF by Azimuthal Walsh Filters, IEEE XPLORE (2015). ISBN: 978-1-4673-7519-1/15
26. I. Bhattacharya, L. Hazra, Azimuthal walsh filters: an interesting tool to produce 2D and 3D light structures, in ed. by A. Martínez-García et al. *Progress in Optomechatronic Technologie*, vol. 233 (Springer Proceedings in Physics, 2019), pp. 11–20
27. A. Sabatyan, J. Rafighdoost, Azimuthal phase-shifted zone plates to produce petal-like beams and ring lattice structures. J. Opt. Soc. Am. A **34**(5), 919–92

Chapter 3
Self-similarity in Transverse Intensity Distributions on the Far-Field Plane of Self-similar Azimuthal Walsh Filters

3.1 Introduction

The fractal or self-similar structures play an important role in describing and understanding a large number of phenomena in several areas of science and technology [1]. In optics, the fractal structure of some optical wave fields and the diffraction pattern generated from various fractal apertures are examples of this category [2, 3]. Fractal zone plates find applications in scientific and technological areas where conventional zone plates have been successfully applied. Modified fractal generalized zone plates extend depth of focus and has been used in spectral domain optical coherence tomography [4]. Fractal zone plates also inspired the invention of other photonic structures [4–6]. Very recent proposals of optical tweezers use phase filters to facilitate the passing of particles in three dimensional structure [7]. Fractal generalized zone plates and spiral fractal zone plates [4] find applications in trapping and optical micro-manipulations [8–14]. Non-uniform distribution of fractal zone plate's focal points along the optical axis could be exploited in the design of multi focal contact lenses. In recent years there is a growing interest in the study and research on quasi-periodic and fractal optical elements. Fibonacci fibre Bragg gratings can transform evanescent wave into propagating waves for far-field superresolution imaging. Fibonacci lenses are being used to generate arrays of optical vortices to trap micro-particle for driving optical pumps [15–24]. Fibonacci diffraction gratings [25, 26] which is an archetypal example of aperiodicity and self-similarity are finding application as image forming devices.

Self-similarity is studied within the distinct groups and subgroups of azimuthal Walsh filters and it has been discussed extensively with illustrations in Sect. 1.6. Self-similarity is also observed in the corresponding transverse intensity distributions of self-similar groups as well as sub-groups of azimuthal Walsh filters. Azimuthal Walsh filters of specific group members are placed on the exit pupil of a rotationally

I. Bhattacharya and L. Hazra, *Azimuthal Walsh Filters*, Progress in Optical Science and Photonics 10,
https://doi.org/10.1007/978-981-15-6099-6_3

symmetric imaging system (refer to Figs. 2.2 and 2.3) and the transverse intensity distributions on the far-field plane are plotted. Groupwise and sub-groupwise self-similar transverse intensity distributions have been presented in the following sections.

3.2 Transverse Intensity Distributions for Zero Order Azimuthal Walsh Filter on the Far-Field Plane

Azimuthal Walsh filter of order zero is placed on the exit pupil plane of a rotationally symmetric imaging system and the normalized transverse intensity distribution on the far-field plane is computed and plotted in Fig. 3.1.

The logarithmic plot of transverse intensity distribution due to zero order azimuthal Walsh function is also computed and shown in Fig. 3.2 where the central maxima is at the origin.

Azimuthal Walsh filter of order zero is called the originating order from which Group 1 has been originated.

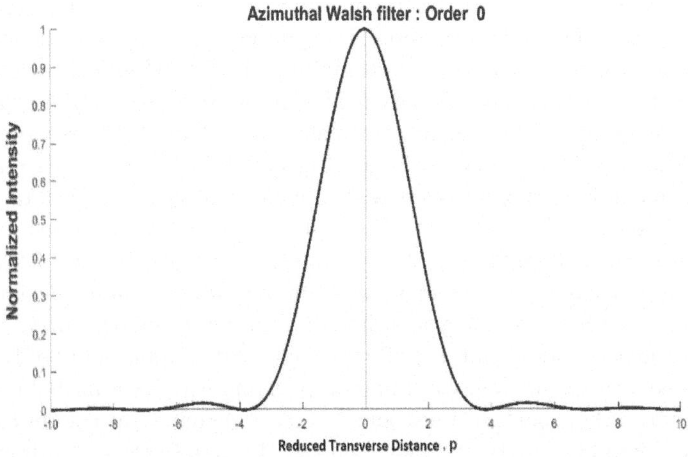

Fig. 3.1 The normalized intensity distribution at far-field plane due to zero order azimuthal Walsh filter on exit pupil

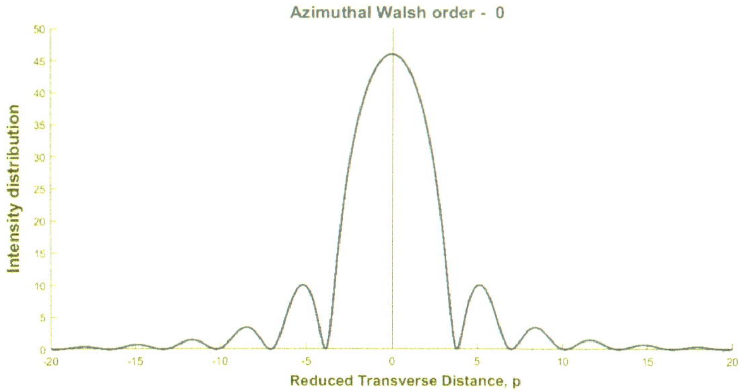

Fig. 3.2 The intensity distribution at far-field plane is plotted on log scale due to zero order azimuthal Walsh filter on exit pupil

3.3 Self-similarity in Far-Field Intensity Distributions for Group I Self-similar Members of Azimuthal Walsh Filters

Azimuthal Walsh filters of Group I consists of orders $\nu = 1, 3, 7, 15, \ldots..$ and the corresponding self-similar transverse intensity distributions on the far-field plane for a particular azimuth, $\zeta = 0$ are illustrated in Fig. 3.3a–d. The intensity at the origin is zero for all members of this group which is the signature of azimuthal Walsh filters.

It is apparent from Fig. 3.3 that the far-field intensity distributions at the transverse plane with members of Group I azimuthal Walsh filters, are similar with central minima associated with first maxima along both sides of the origin followed by second, third, fourth, ... maxima of diminishing amplitudes. For higher orders of Group I self-similar members, the width of the central minima increases and the sharpness of first maxima increases.

3.4 Self-similarity in Far-Field Intensity Distributions for Group IIA Self-similar Members of Azimuthal Walsh Filters

The members of Group IIA consists of azimuthal Walsh filters of orders, $\nu = 2, 4, 8,$ ...... and their corresponding transverse intensity distributions at the far-field plane for azimuth $\zeta = 0$ are shown in Fig. 3.4a–c) respectively.

Figure 3.4 shows that the transverse intensity distributions at the far-field plane for self-similar members of Group IIA azimuthal Walsh filters are similar to each other. The patterns are having central minima associated with sharp first maxima along both sides of the origin followed by successive maxima of diminishing amplitudes.

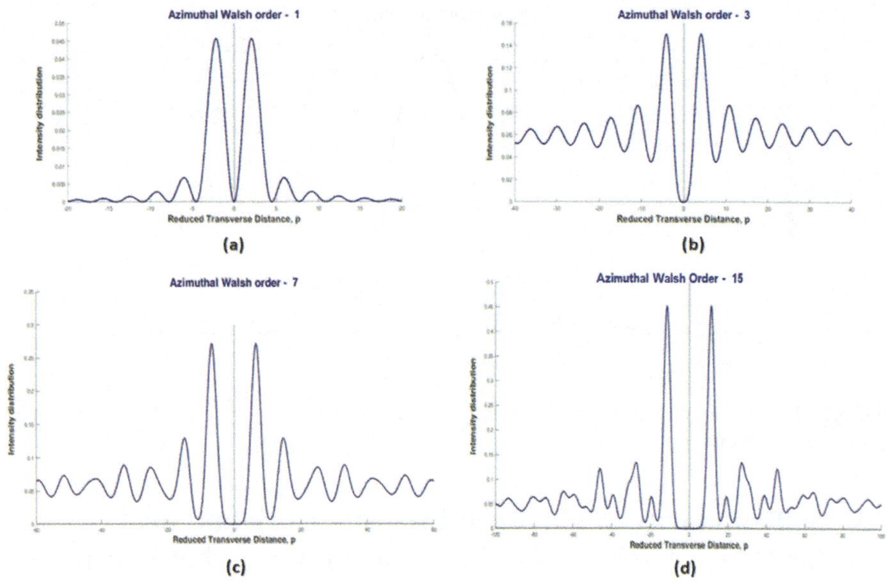

Fig. 3.3 **a–d** Transverse intensity distributions at far-field plane of azimuthal Walsh filters of Group I members of orders, $\nu = 1, 3, 7$ and 15 for azimuth, $\zeta = 0$

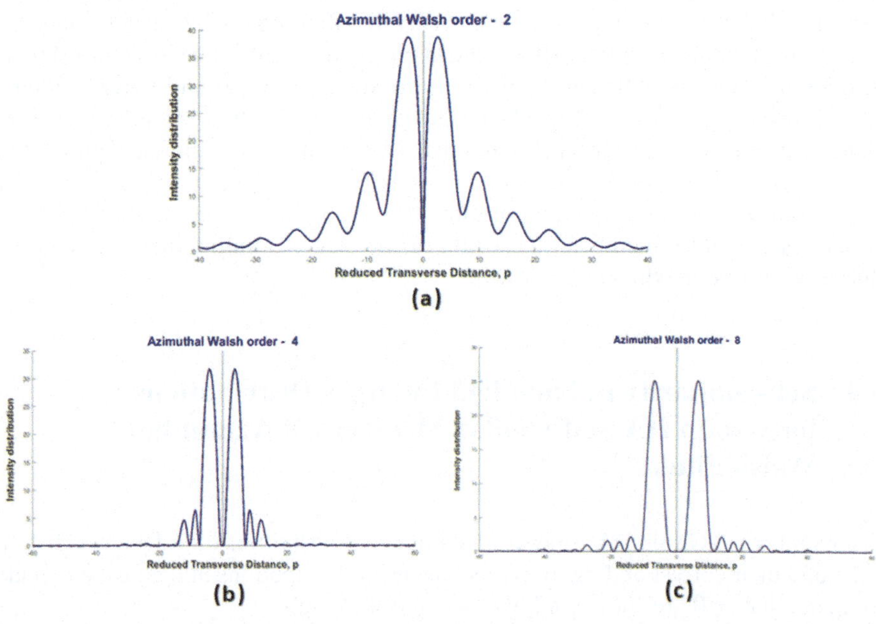

Fig. 3.4 **a–c** Transverse intensity distributions at far-field plane of azimuthal Walsh filters of Group IIA members of orders $\nu = 2, 4$ and 8 for azimuth, $\zeta = 0$

For higher orders, the width of the central minima increases and the sharpness of first maxima increases as exhibited by Group I members.

3.5 Self-similarity in Far-Field Intensity Distributions for Group IIB Self-similar Members of Azimuthal Walsh Filters

Group IIB consists of azimuthal Walsh filters of orders 6, 12, 24, 48 .... as shown in Table 1.4. The transverse intensity distributions at far-field plane due to azimuthal Walsh filters of orders 6 and 12 from the self-similar Group IIB for azimuth $\zeta = 0$ are illustrated in Fig. 3.5a, b respectively.

From Fig. 3.5 it is apparent that for the Group IIA members, the transverse intensity distributions at far-field plane show the central minima along with first maxima immediately followed by the second one along both sides of the origin. The intensity reduced to zero for both orders as one move away from the origin of the transverse far-field plane.

3.6 Self-similarity in Far-Field Intensity Distributions for Group IIIA Self-similar Members of Azimuthal Walsh Filters

Subgroup IIIAA consists of azimuthal Walsh filters of orders 5, 11, 23, 47, ... as shown in Table 1.5. The transverse intensity distributions due to azimuthal Walsh filters for orders $\nu = 5$ and 11 are shown in Fig. 3.6a, b respectively. The patterns show amazing resemblance with central minima followed by 1st, 2nd, 3rd, ... maxima of diminishing amplitudes. Intensity distribution in both cases dies down as one move away from the origin at the centre of the far-field plane.

It is observed from Figs. 3.3, 3.4, 3.4 and 3.6 that the transverse intensity distributions at the far-field plane due to each group and subgroup of azimuthal Walsh filters, are showing self-similarity as a consequence of self-similarity existing within the members of corresponding group and subgroup as depicted in Tables 1.2, 1.3, 1.4 and 1.5. It is also observed that as the order, ν increases within each subgroup, the maxima of the intensity distributions become sharper with increasing ripples of diminishing amplitudes in the sidelobes.

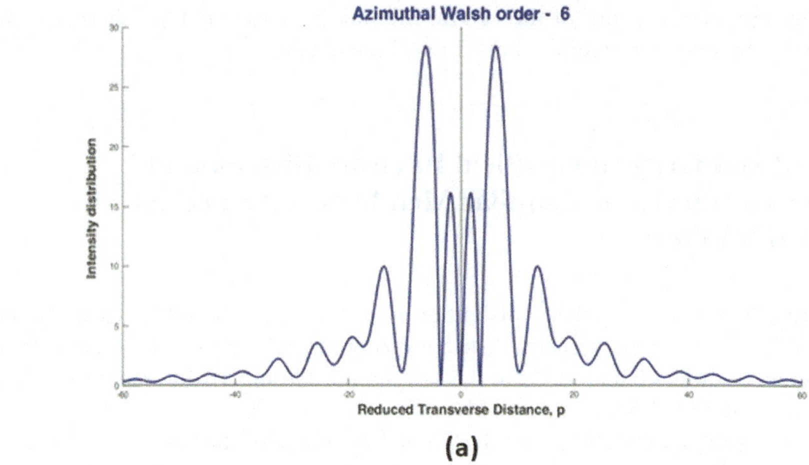

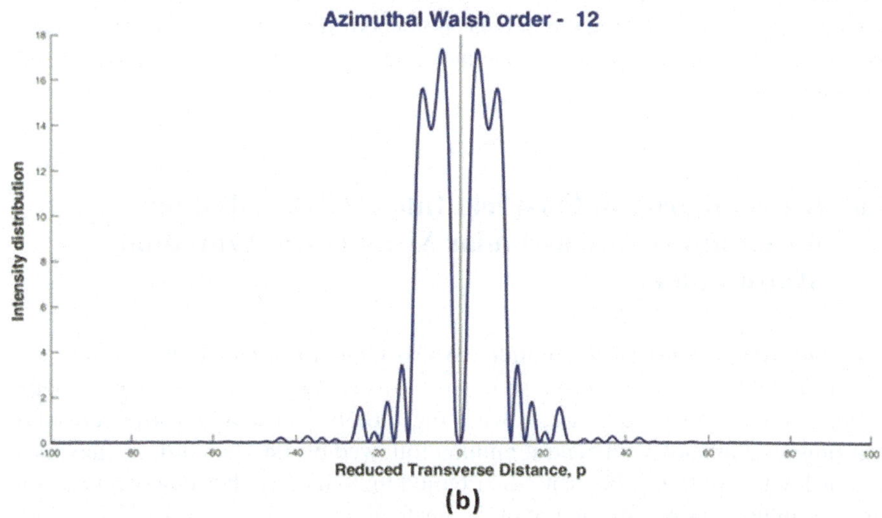

Fig. 3.5 **a, b** Transverse intensity distributions at far-field plane of azimuthal Walsh filters of self-similar members of Group IIA for orders $\nu = 6$ and 12 for $\zeta = 0$

3.7 Rotational Self-similarity Observed in 2D Transverse Intensity Distributions at Far-Field Plane for Adjacent Orders of Azimuthal Walsh Filters

In addition to the inherent self-similarity exhibited like radial and annular varieties, 2D intensity distributions at transverse far-field plane due to azimuthal Walsh filters are observed to possess a unique rotational self-similarity exhibited among adjacent orders. The 2D intensity distributions due to a succeeding ordered azimuthal Walsh filter can be retrieved from the preceding order by just rotating it by some specified

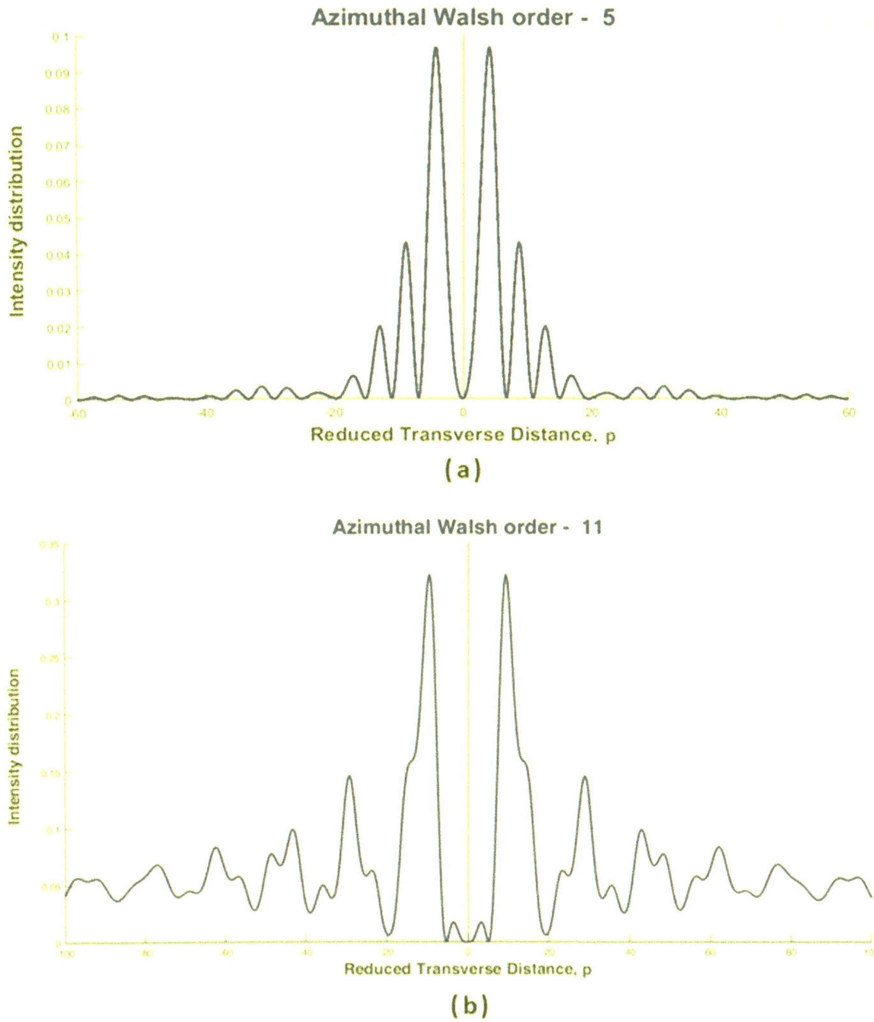

Fig. 3.6 **a**, **b** Transverse intensity distributions at far-field plane of self-similar azimuthal Walsh filters of Group IIA of orders $\nu = 5$ and 11, for azimuth $\zeta = 0$

angle like π, $\pi/2$, $\pi/4$, $\pi/8$ ... and the angle of rotation depends upon the order ν of the azimuthal Walsh filter under consideration. The rotational self-similarity observed in transverse 2D intensity distributions at far-field plane for adjacent orders with the respective angle rotation for orders $\nu = 0, 1, \ldots, 18$ is discussed in detail in [27].

References

1. H. Takayasu, *Fractals in Physical Science* (Manchester University, Manchester, 1990)
2. C. Allain, M. Cloitre, Optical diffraction on fractals. Phys. Rev. B **33**, 3566 (1986)
3. J. Uozumi, T. Asakura, Fractal Optics, in *Current Trends in Optics*, ed. by J.C. Dainty (Academic Press, Cambridge, London, 1994), pp. 189–196
4. Q.Q. Zhang, J.G. Wang et al, A modified fractal zone plate with extended depth of focus in spectral domain optical coherence tomography. J. Opt. **13**(5), 055301,6 (2011)
5. S.H. Tao, X.C. Yuan et al., Sequence of focused optical vortices generated by a spiral fractal zone plate. Appl. Phys. Lett. **89**(3), 031105 (2006)
6. J.A. Monsoriu, W.D. Furlan, P. Andreas, J. Lancis, Fractal conical lenses. Opt. Exp. **14**(20), 9077–9082 (2006)
7. V. Ferrando, A. Calatayud, F. Gimenez, W.D. Furlan, J.A. Monsoriu, Cantor dust zone plates. Opt. Exp. **21**(3), 2701–2706 (2013)
8. H. Melville, G.F. Milne, Optical trapping of three-dimensional structures using dynamic holograms. Opt. Exp. **11**(26), 3562–3567 (2003)
9. E. Schonbrun, C. Rinzler, K.B. Crozier, Microfabricated water immersion zone plate optical tweezer. Appl. Phys. Lett. **92**, 071112 (2008)
10. A. Ashkin, J.M. Dziedzic et al., Observation of a single-beam gradient force optical trap for dielectric particles. Opt. Lett. **11**(5), 288–290 (1986)
11. K.C. Neumann, S.M. Block, Optical trapping. Rev. Sci. Instrum. **75**(9), 2787–2809 (2004)
12. K. Dholakia, T. Cizmar, Shaping the future manipulation. Nat. Photonics **5**, 335–342 (2011)
13. J.E. Molloy, M.J. Padgett, Lights, action: optical tweezers. Contemp. Phys. **43**(4), 241–258 (2002)
14. D.G. Grier, A revolution in optical manipulation. Nature (London) **424**(6950), 810–816 (2003)
15. J. Zhang, Y. Cao, J. Zheng, Fibonacci quasi-periodic superstructure fiber Bragg gratings. Optik **121**(5), 417–421 (2010)
16. K. Wu, G.P. Wang, One-dimensional Fibonacci grating for far-field super-resolution imaging. Opt. Lett. **38**(12), 2032–2034 (2013)
17. A. Calatayud, L. Remon, W.D. Furlan, J.A. Monsoriu, Twin axial vortices generated by Fibinacci lenses. Opt. Exp. **21**(8), 10234–10239 (2013)
18. J.A. Monsoriu, C.J. Zapata-Rodriguez, W.D. Furlan, Fractal axicons. Opt. Commun. **263**, 1–5 (2006)
19. R. Verma, V. Banerjee, P. Senthilkumaran, Redundancy in Cantor diffractals. Opt. Commun. **263**, 1–5 (2006)
20. R. Verma, M.K. Sharma, V. Banerjee, P. Senthilkumaran, Robustness of Cantor diffractals. Opt. Exp. **21**(7), 7951–7956 (2013)
21. W. Gallermann, M. Kohmoto, B. Sutherland, P.C. Taylor, Localization of light waves in Fibonacci dielectric multilayers. Phys. Rev. Lett. **72**(5), 633–636 (1994)
22. X. Yang, Y. Liu, X. Fu, Transmission properties of light through the Fibonacci-class multilayers. Phys. Rev. B **59**(7), 4545–4548 (1999)
23. N.V. Grushina, P.V. Korolenko, S.N. Markova, Special features of the diffraction of light on optical Fibonacci gratings. Moscow Univ. Phys. Bull. **63**(2), 123–126 (2008)
24. N. Gao, Y. Zhang, C. Xie, Circular Fibonacci gratings. Appl. Opt. **50**(31), G142–G148 (2011)
25. R. Verma, V. Banerjee, P. Senthilkumaran, Fractal signatures in the aperiodic Fibonacci grating. Opt. Lett. **39**(9), 2557–2560 (2014)
26. R. Verma, M.K. Sharma, P. Senthilkumaran, V. Banerjee, Analysis of Fibonacci gratings and their diffraction patterns. J. Opt. Soc. Am. A **31**(7), 1473–1480 (2014)
27. I. Bhattacharya, Self-similarity in azimuthal Walsh filters and corresponding far-field diffraction characteristics: a unique study to control tightly focused fields and coupling of light into metamaterials, plasmonic structure and waveguides, in Proceedings of SPIE 11257, Plasmonics in Biology and Medicine (vol. XVII), 2020, pp. 1125717:1-14

Chapter 4
Transverse Intensity Distribution in the Far-Field Region of Azimuthal Walsh Filters

4.1 Introduction

Azimuthal Walsh filters $W_\nu(\theta)$, for order $\nu = 0, 1, 2, \ldots$ are placed on the exit pupil of a rotationally symmetric imaging system (Fig. 4.1) and the normalized intensity at the point Q on an axially shifted image plane is computed by adding a defocus aberration term using another alternative method. The plane where Q located is shifted longitudinally from the paraxial focal plane or image plane by an actual longitudinal shift ΔZ. a represents the reduced axial coordinate or distance as expressed by Eq. (4.7). The normalized intensity $I_N(p, \zeta; W_{20})$ is dependent on the actual longitudinal shift ΔZ and hence on the reduced axial coordinate a. Normalized intensities are computed for different values of reduced axial distance a. Plots of normalized intensity with reduced transverse distance are shown for all orders of azimuthal Walsh filters from 0 to 7 and are presented in the next sections. The origin of the intensity distribution curves represents the location of the paraxial focal plane or far-field plane (Fig. 4.1) and the positive and negative values of a represent axial shifts on either side of the paraxial focal plane or far-field plane. The analytical method and the illustrative results with discussions are shown in the next sections.

4.2 Analytical Formulation of Intensity Distribution on Axially Shifted Image Planes

The transverse image plane of a rotationally symmetric system is shifted longitudinally from the paraxial focal plane by an amount ΔZ is shown in Fig. 4.1.

I. Bhattacharya and L. Hazra, *Azimuthal Walsh Filters*, Progress in Optical Science and Photonics 10,
https://doi.org/10.1007/978-981-15-6099-6_4

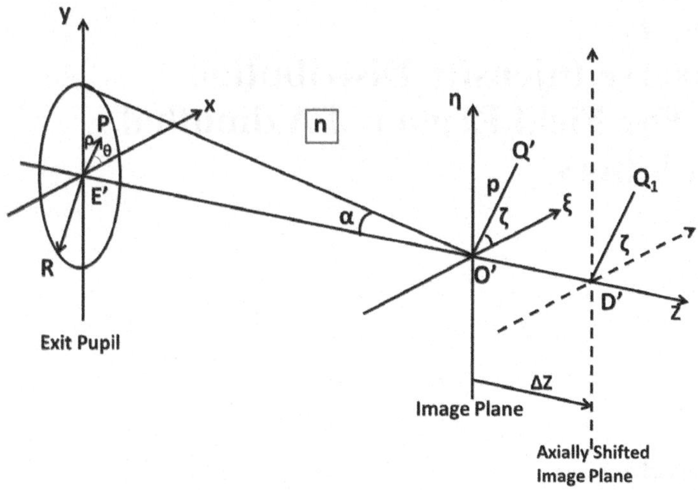

Fig. 4.1 Exit pupil, image plane and an axially shifted image plane in the image space of a rotationally symmetric imaging system

Considering the case of azimuthal Walsh filter, $W_\nu(\theta)$ of order ν placed at the exit pupil plane of the imaging system,

$$f(\theta) = W_\nu(\theta) = \sum_{m=1}^{M} e^{ik\psi_m} \mathfrak{B}_m(\theta) \tag{4.1}$$

where $\mathfrak{B}_m(\theta)$ are azimuthal Walsh Block functions [1] defined as,

$$\begin{aligned} \mathfrak{B}_m(\theta) &= 1; \quad \theta_{m-1} \le \theta \le \theta_m \\ &= 0; \quad \text{otherwise} \end{aligned} \tag{4.2}$$

where

$$\theta_{m-1} = \left(\frac{m-1}{M}\right)2\pi \quad \text{and} \quad \theta_m = \left(\frac{m}{M}\right)2\pi \tag{4.3}$$

with $M = 2^\alpha$ is the largest power of 2 that just exceeds ν. Over the mth sector of $W_\nu(\theta)$, value of $k\psi_m$ is 0 or π, if $W_\nu(\theta) = +1$ or -1 respectively.

The complex amplitude distribution at a transverse plane shifted from the paraxial focal plane by an amount ΔZ can be obtained by adding a defocus aberration term W (r) in the Exp. (4.1), for $f(r,\theta) = \exp[i\,k\,\{W_\nu(\theta) + W(r)\}]$ and we can write:

$$f(r,\theta) = \exp[i\,k\,\{W_\nu(\theta) + W(r)\}] \tag{4.4}$$

where $W(r)$ is the wave aberration function defined as

$$W(r) = W_{20}r^2 \tag{4.5}$$

A value of

$$kW_{20} = a\pi \tag{4.6}$$

corresponds to the reduced axial coordinate a, given by:

$$a = \frac{1}{n\lambda}(n\sin\alpha)^2\Delta Z \tag{4.7}$$

where n represents the refractive index of the image space and $(n\sin\alpha)$ is the image space numerical aperture.

The shift of image plane from paraxial focal plane by an amount ΔZ causes defocus or 'Depth of Focus' aberration which is a measure of asphericity of the image forming wavefront as shown in Fig. 4.2.

The complex amplitude distribution on the axially shifted image plane is:

$$A(p,\zeta;W_{20}) = \int_0^1\int_0^{2\pi} \exp\left[ik\left\{W_\nu(\theta) + W_{20}r^2\right\}\right]\exp[ipr\cos\omega]r\,dr\,d\theta \tag{4.8}$$

where $\omega = (\theta - \zeta)$ and $R = 1$.

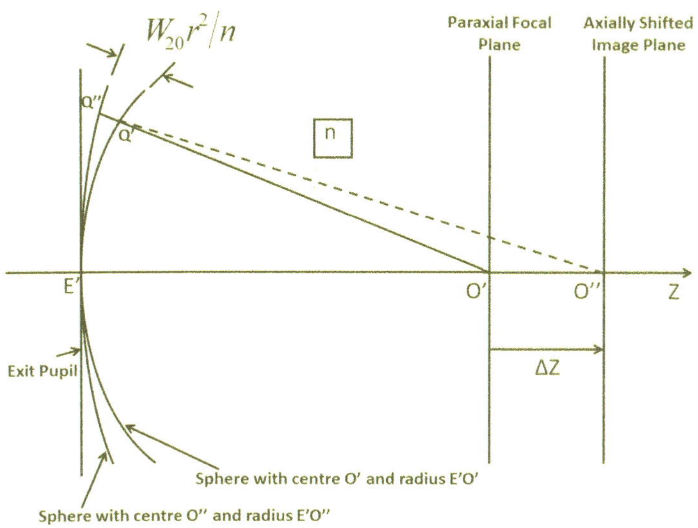

Fig. 4.2 Shift of image plane from paraxial focal plane by ΔZ causes defocus or 'Depth of Focus'

At the origin,

$$A(0,0;W_{20})=\int_0^1\int_0^{2\pi}\exp\left[ik\left\{W_\nu(\theta)+W_{20}r^2\right\}\right]rdrd\theta \tag{4.9}$$

i.e.,

$$A(0,0;W_{20})=\int_0^{2\pi}e^{ikW_\nu(\theta)}d\theta\int_0^1 e^{ikW_{20}r^2}rdr=A_1A_2 \tag{4.10}$$

where

$$A_1=\int_0^{2\pi}e^{ikW_\nu(\theta)}d\theta \tag{4.11}$$

and

$$A_2=\int_0^1 e^{ikW_{20}r^2}rdr \tag{4.12}$$

To evaluate A_1, the value of $W_\nu(\theta)$ is to be substituted from Eq. (4.1) to Exp. (4.11) and we have:

$$A_1=\int_0^{2\pi}e^{ikW_\nu(\theta)}d\theta=\sum_{m=1}^{M}e^{ik\psi_m}\int_0^{2\pi}\mathfrak{B}_m(\theta)d\theta \tag{4.13}$$

To evaluate A_2, we will substitute $r^2 = t$ so that $2rdr = dt$ and substituting in Eq. (4.12) we have:

$$A_2=\int_0^1 e^{ikW_{20}r^2}rdr=\frac{1}{2}\int_0^1 e^{ikW_{20}t}dt\frac{1}{2ikW_{20}}\left[e^{ikW_{20}}-1\right] \tag{4.14}$$

Hence,

$$A(0,0;W_{20}) = A_1A_2 = \left\{\sum_{m=1}^{M} e^{ik\psi_m}\int_{0}^{2\pi}\mathfrak{B}_m(\theta)d\theta\right\}\mathrm{X}\left[\frac{1}{2ikW_{20}}\left[e^{ikW_{20}}-1\right]\right] \tag{4.15}$$

The intensity distribution at origin of the shifted image plane is,

$$I(0,0;W_{20}) = |A(0,0;W_{20})|^2 \tag{4.16}$$

At any point (p, ζ) on the image plane, the complex amplitude distribution can be written from Eq. (4.8) by substituting the value of $W_\nu(\theta)$ from Eq. (4.1) and W(r) from Eq. (4.5) as:

$$A(p,\zeta;W_{20}) = \sum_{m=1}^{M} e^{ik\psi_m}\int_{0}^{1}\int_{0}^{2\pi}\mathfrak{B}_m(\theta)\exp\left[ikW_{20}r^2\right]\exp[ipr\cos\omega]rdrd\theta \tag{4.17}$$

Expressing,

$$\tau(r,\theta) = \exp\left[ikW_{20}r^2\right]\exp[ipr\cos\omega] \tag{4.18}$$

Equation (4.17) can be further simplified as,

$$A(p,\zeta;W_{20}) = \sum_{m=1}^{M} e^{ik\psi_m}\int_{0}^{1}\int_{0}^{2\pi}\mathfrak{B}_m^M(\theta)\tau(r,\theta)rdrd\theta \tag{4.19}$$

4.3 Synthesis of Azimuthal Walsh Filters Using Azimuthal Walsh Block Functions

The synthesis of azimuthal Walsh filters $W_\nu(\theta)$, for order $\nu = 0,1,2,3$ from azimuthal Walsh Block functions, $\mathfrak{B}_m(\theta)$ and the derivation of corresponding complex amplitude and normalized intensity distributions at the shifted image planes is shown below:

Case 1: For zero order azimuthal Walsh filter, $W_0(\theta)$ when $\nu = 0$

For Zero order azimuthal Walsh filter, $\nu = 0$, m = 1, M = 1, $\theta_{m-1} = 0$, $\theta_m = 2\pi$, $k\psi_1 = 0$ (Fig. 4.3),

Fig. 4.3 Synthesis of zero order azimuthal Walsh function, $W_0(\theta)$ using azimuthal Walsh Block function $\mathfrak{B}_1^i(\theta)$, where $i = 1$

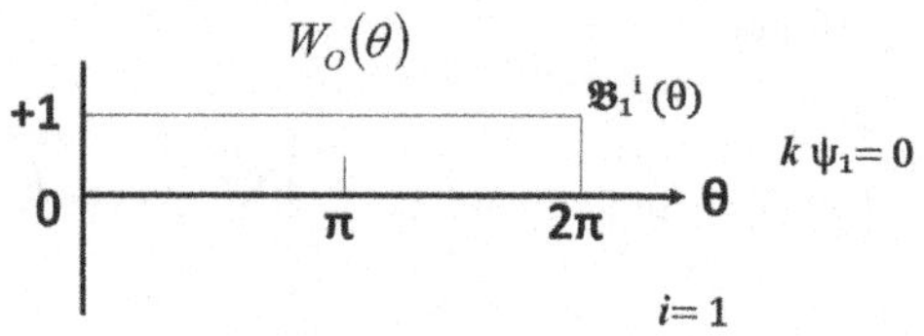

$$W_\nu(\theta)|_{\nu=0} = W_0(\theta) = e^{ik\psi_1}\mathfrak{B}_1^1(\theta) = \mathfrak{B}_1^1(\theta) \tag{4.20}$$

where

$$\mathfrak{B}_1^1(\theta) = 1, 0 \le \theta \le 2\pi \tag{4.21}$$

Amplitude distribution at shifted image plane due to zero order azimuthal Walsh filter at exit pupil can be computed from Eq. (4.19) as:

$$A(p, \zeta; W_{20})|_{\nu=0} = \int_0^1 \int_0^{2\pi} \exp\left[ikW_{20}r^2\right] \exp[ipr\cos\omega] r dr d\theta \tag{4.22}$$

$$=> A(p, \zeta; W_{20})|_{\nu=0} = \int_0^1 \int_0^{2\pi} \tau(r, \theta) r dr d\theta \tag{4.23}$$

Normalized Intensity distribution at the plane longitudinally shifted by ΔZ can be written as:

$$I_N(p, \zeta; W_{20})|_{\nu=0} = \frac{I(p, \zeta; W_{20})|_{\nu=0}}{I(0, 0; W_{20})|_{\nu=0}} = \frac{\left|A(p, \zeta; W_{20})|_{\nu=0}\right|^2}{|A(0, 0; W_{20})|_{\nu=0}|^2} \tag{4.24}$$

Case 2: For first order azimuthal Walsh filter, $W_1(\theta)$
For first order azimuthal Walsh filter, $\nu = 1$; $M = 2^\alpha|_{\alpha=1} = 2$; m = 1, 2; and we can write,

$$W_\nu(\theta)|_{\nu=1} = W_1(\theta) = \sum_{m=1}^{2} e^{ik\psi_m}\mathfrak{B}_m^M(\theta) \tag{4.25}$$

$$=> W_1(\theta) = e^{ik\psi_1}\mathfrak{B}_1^2(\theta) + e^{ik\psi_2}\mathfrak{B}_2^2(\theta) \tag{4.26}$$

$W_1(\theta)$ is expressed as summation of two Walsh Block functions, $\mathfrak{B}_1^2(\theta)$ and $\mathfrak{B}_2^2(\theta)$ defined as:

$$\mathfrak{B}_1^2(\theta) = 1; \quad 0 \le \theta \le \pi = 0; \quad \pi \le \theta \le 2\pi \tag{4.27}$$

$$\mathfrak{B}_2^2(\theta) = 1; \quad \pi \le \theta \le 2\pi = 0; \quad 0 \le \theta \le \pi \tag{4.28}$$

Also, over the m = 1 sector, the value of $W_1(\theta)$ is +1, so the value of $k\psi_1 = 0$ and over the m = 2 sector, the value of $W_1(\theta)$ is −1 and hence $k\psi_2 = \pi$.

Substituting these values of $k\psi_1, k\psi_2$ and $\mathfrak{B}_1^2(\theta)$ and $\mathfrak{B}_2^2(\theta)$ from Eqs. (4.27) and (4.28) in Eq. (4.26) we can write,

$$W_1(\theta) = \mathfrak{B}_1^2(\theta) - \mathfrak{B}_2^2(\theta) \tag{4.29}$$

Synthesis of first order Walsh filter $W_1(\theta)$ using Walsh Block functions $\mathfrak{B}_1^2(\theta)$, $\mathfrak{B}_2^2(\theta)$ is depicted schematically in Fig. 4.4.

The complex amplitude distribution at shifted image plane due to first order azimuthal Walsh filter, $W_1(\theta)$ on exit pupil can be written as,

$$=> A(p, \zeta; W_{20})|_{\upsilon=1} = \int_0^1 \int_0^{\pi} \tau(r, \theta) r dr d\theta - \int_0^1 \int_{\pi}^{2\pi} \tau(r, \theta) r dr d\theta \tag{4.30}$$

The normalized Intensity distribution is:

$$I_N(p, \zeta; W_{20})|_{\upsilon=1} = \frac{I(p, \zeta; W_{20})|_{\upsilon=1}}{I(0, 0; W_{20})|_{\upsilon=1}} = \frac{|A(p, \zeta; W_{20})|_{\upsilon=1}|^2}{|A(0, 0; W_{20})|_{\upsilon=1}|^2} \tag{4.31}$$

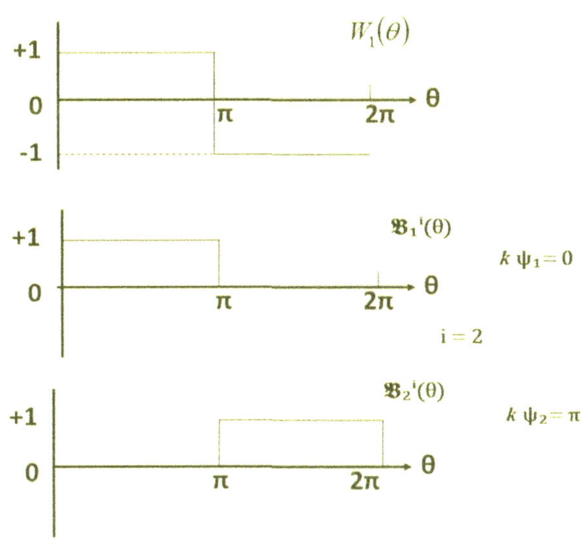

Fig. 4.4 Synthesis of first order azimuthal Walsh function, $W_1(\theta)$ using Walsh Block functions, $\mathfrak{B}_1^i(\theta)$ and $\mathfrak{B}_2^i(\theta)$, where $i = 2$

Case 3: Second order azimuthal Walsh filter, $W_2(\theta)$
For second order azimuthal Walsh filter, $\nu = 2$; $M = 2^{\alpha}|_{\alpha=2} = 4$; m = 1, 2, 3, 4; and hence,

$$W_\nu(\theta)|_{\nu=2} = W_2(\theta) = \sum_{m=1}^{4} e^{ik\psi_m} \mathfrak{B}_m^M(\theta)$$

$$W_2(\theta) = e^{ik\psi_1}\mathfrak{B}_1^4(\theta) + e^{ik\psi_2}\mathfrak{B}_2^4(\theta) + e^{ik\psi_3}\mathfrak{B}_3^4(\theta) + e^{ik\psi_4}\mathfrak{B}_4^4(\theta) \tag{4.32}$$

$W_2(\theta)$ is expressed as a summation of four azimuthal Walsh Block functions, $\mathfrak{B}_1^4(\theta)$, $\mathfrak{B}_2^4(\theta)$, $\mathfrak{B}_3^4(\theta)$ and $\mathfrak{B}_4^4(\theta)$ defined by:

$$\begin{aligned} \mathfrak{B}_m^M(\theta) &= +1, \ \frac{(m-1)}{M}2\pi \le \theta \le \frac{m}{M}2\pi \\ &= 0, \quad \text{otherwise} \end{aligned} \tag{4.33}$$

Also, over m = 1 sector, the value of $W_2(\theta)$ is +1, so the value of $k\psi_1 = 0$; over m = 2 sector, $W_2(\theta)$ is −1 and $k\psi_2 = \pi$; over m = 3 sector, the value of $W_2(\theta)$ is −1 and $k\psi_3 = \pi$; and over m = 4 sector, $W_2(\theta)$ is +1 and $k\psi_4 = 0$.

Substituting the values of $k\psi_1$, $k\psi_2$, $k\psi_3$, $k\psi_4$ and $\mathfrak{B}_m^M(\theta)$'s evaluated from Eq. (4.33) in Eq. (4.32) and we can write:

$$W_2(\theta) = \mathfrak{B}_1^4(\theta) - \mathfrak{B}_2^4(\theta) - \mathfrak{B}_3^4(\theta) + \mathfrak{B}_4^4(\theta) \tag{4.34}$$

The synthesis of $W_2(\theta)$ from azimuthal Walsh Block functions $\mathfrak{B}_m^M(\theta)$ where M = 4 and m = 1, 2, 3, 4 is schematically represented in Fig. 4.5. The complex amplitude distribution at the shifted image plane is,

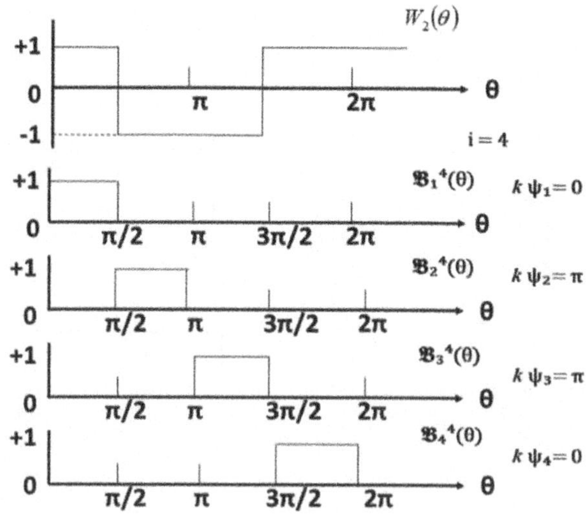

Fig. 4.5 Synthesis of second order azimuthal Walsh function, $W_2(\theta)$ using Walsh Block functions $\mathfrak{B}_1^i(\theta)$, $\mathfrak{B}_2^i(\theta)$, $\mathfrak{B}_3^i(\theta)$, $\mathfrak{B}_4^i(\theta)$, where $i = 4$

$$A(p,\zeta;W_{20})|_{\nu=2}=\int_0^1\int_0^{\pi/2}\tau(r,\theta)rdrd\theta-\int_0^1\int_{\pi/2}^{\pi}\tau(r,\theta)rdrd\theta$$
$$-\int_0^1\int_{\pi}^{3\pi/2}\tau(r,\theta)rdrd\theta+\int_0^1\int_{3\pi/2}^{2\pi}\tau(r,\theta)rdrd\theta \quad (4.35)$$

The normalized intensity distribution at that plane is:

$$I_N(p,\zeta;W_{20})|_{\nu=2}=\frac{I(p,\zeta;W_{20})|_{\nu=2}}{I(0,0;W_{20})|_{\nu=2}}=\frac{|A(p,\zeta;W_{20})|_{\nu=2}|^2}{|A(0,0;W_{20})|_{\nu=2}|^2} \quad (4.36)$$

Case 4: Third order azimuthal Walsh filter, $W_3(\theta)$

Third order azimuthal Walsh filter can be synthesized by defining the parameters, $\nu=3$; $M=2^{\alpha}|_{\alpha=2}=4$; m = 1, 2, 3, 4; and hence:

$$W_\nu(\theta)|_{\nu=3}=W_3(\theta)=\sum_{m=1}^{4}e^{ik\psi_m}\mathfrak{B}_m^M(\theta)$$
$$=>W_3(\theta)=e^{ik\psi_1}\mathfrak{B}_1^4(\theta)+e^{ik\psi_2}\mathfrak{B}_2^4(\theta)+e^{ik\psi_3}\mathfrak{B}_3^4(\theta)+e^{ik\psi_4}\mathfrak{B}_4^4(\theta) \quad (4.37)$$

$W_3(\theta)$ is expressed as summation of four azimuthal Walsh Block functions, $\mathfrak{B}_1^4(\theta)$, $\mathfrak{B}_2^4(\theta)$, $\mathfrak{B}_3^4(\theta)$ and $\mathfrak{B}_4^4(\theta)$ as shown in Exp. (4.37).

In this case, over the first sector, the value of $W_3(\theta)$ is $+1$, so the value of $k\psi_1=0$; over the second sector, the value of $W_3(\theta)$ is -1 and hence $k\psi_2=\pi$; over third sector, the value of $W_3(\theta)$ is $+1$ and hence $k\psi_3=0$; and over fourth sector, the value of $W_3(\theta)$ is -1 resulting the value of $k\psi_4=\pi$.

Substituting the values of $k\psi_1$, $k\psi_2$, $k\psi_3$, $k\psi_4$ and $\mathfrak{B}_m^M(\theta)$'s evaluated from Eq. (4.33), into Eq. (4.37), we can write:

$$W_3(\theta)=\mathfrak{B}_1^4(\theta)-\mathfrak{B}_2^4(\theta)+\mathfrak{B}_3^4(\theta)-\mathfrak{B}_4^4(\theta) \quad (4.38)$$

The synthesis of $W_3(\theta)$ from azimuthal Walsh Block functions $\mathfrak{B}_m^M(\theta)$ where M = 4 and m = 1, 2, 3, 4 is schematically represented in Fig. 4.6.

Therefore, complex amplitude distribution at the shifted image plane due to $W_3(\theta)$ on exit pupil plane can be expressed as,

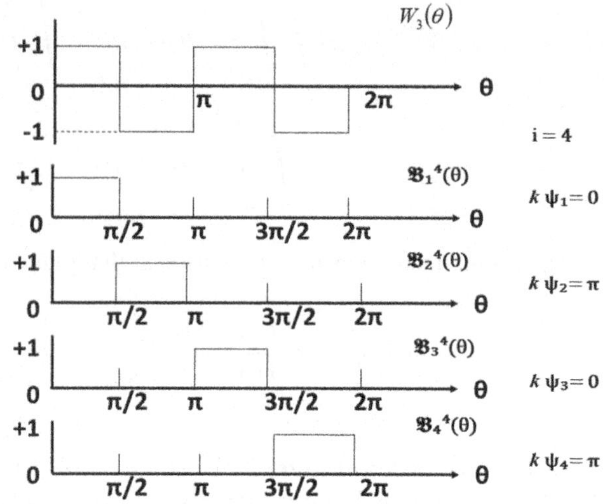

Fig. 4.6 Synthesis of third order azimuthal Walsh function $W_3(\theta)$ using azimuthal Walsh Block functions $\mathfrak{B}_1^i(\theta), \mathfrak{B}_2^i(\theta), \mathfrak{B}_3^i(\theta), \mathfrak{B}_4^i(\theta)$, where $i = 4$

$$A(p, \zeta; W_{20})|_{\nu=3} = \int_0^1 \int_0^{\pi/2} \tau(r, \theta) r dr d\theta - \int_0^1 \int_{\pi/2}^{\pi} \tau(r, \theta) r dr d\theta + \int_0^1 \int_{\pi}^{3\pi/2} \tau(r, \theta) r dr d\theta - \int_0^1 \int_{3\pi/2}^{2\pi} \tau(r, \theta) r dr d\theta \quad (4.39)$$

The corresponding normalized intensity distribution at that plane is:

$$I_N(p, \zeta; W_{20})|_{\nu=3} = \frac{I(p, \zeta; W_{20})|_{\nu=3}}{I(0, 0; W_{20})|_{\nu=3}} = \frac{|A(p, \zeta; W_{20})|_{\nu=3}|^2}{|A(0, 0; W_{20})|_{\nu=3}|^2} \quad (4.40)$$

In this section, azimuthal Walsh filters $W_\nu(\theta)$ of orders $\nu = 0, 1, 2, 3$ have been synthesized using Walsh Block functions $\mathfrak{B}_m^M(\theta)$ and shown both analytically as well as pictorially with $M = 2^\alpha$, the largest power of 2 that just exceeds ν and $m = 1, \ldots, M$, the number of sectors by which the filter is constituted.

4.4 Evaluation of Integral Using the Concept of Concentric Equal Area Zones of Azimuthal Walsh Filters

To evaluate the integral in Eq. (4.22) namely,

$$A(p,\zeta;W_{20})|_{v=0} = \int_0^1 \int_0^{2\pi} \exp\left[ikW_{20}r^2\right]\exp[ipr\cos\omega]rdrd\theta \tag{4.22}$$

where $\omega = \theta - \zeta$, the following method of separation of integrands of variables r and θ has been adopted.

Re-arranging Eq. (4.22), we can write:

$$A(p,\zeta) = \int_0^1 \exp\left[ikW_{20}r^2\right]\left\{\int_0^{2\pi} \exp[ipr\cos(\theta-\zeta)]d\theta\right\}rdr \tag{4.41}$$

Defining $r^2 = t$, so that $2rdr = dt$ and substituting in Eq. (4.41) we have:

$$A(p,\zeta) = \frac{1}{2}\int_0^1 \exp[ikW_{20}t]\left\{\int_0^{2\pi} \exp\left[ip\sqrt{t}\cos(\theta-\zeta)\right]d\theta\right\}dt \tag{4.42}$$

Defining I as:

$$I = \exp(ig(t)) = \int_0^{2\pi} \exp\left[ip\sqrt{t}\cos(\theta-\zeta)\right]d\theta \tag{4.43}$$

and putting in Eq. (4.42), the complex amplitude distribution at the shifted image plane can be further simplified as:

$$A(p,\zeta) = \frac{1}{2}\int_0^1 \exp[i\{kW_{20}t + g(t)\}]dt \tag{4.44}$$

To evaluate the integral I in Eq. (4.43), a set of values of reduced coordinates at the image plane (p_m, ζ_n) has been defined and for these values $I = I_1$ takes the form as:

$$I_1 = \exp(ig(t)) = \int_0^{2\pi} \exp\left[ip_m\sqrt{t}\cos(\theta-\zeta_n)\right]d\theta \tag{4.45}$$

For better accuracy of results, each of the M zones of the pupil will be subdivided into $\mathcal{L}$ number of equal area subzones such that the whole exit pupil has $= M\mathcal{L}$ number of equal area concentric zones. For each of these sub zones having outer and inner radii r_1 and r_2, the value of r_ℓ can be evaluated as:

$$\mathrm{r}_\ell = r_1 + \ell . \frac{r_2 - r_1}{\mathcal{L}} \tag{4.46}$$

where = 0, 1,…, $\mathcal{L}$.

We have divided the exit pupil zone in = 2^9 = 512 numbers and the integration has been carried out for 512 values of $\mathrm{t}_\ell = \sqrt{\mathrm{r}_\ell}$ and the result has been stored within,

$$(I_1)_{m,n,l} = \int_0^{2\pi} exp[i\, \hbar(p_m, \zeta_n, t_\ell, \theta)]d\theta \tag{4.47}$$

The integration in Exp. (4.47) is reduced to a single variable integrand, θ and it has been evaluated within the specified domain of (0, 2π).

Azimuthal Walsh filters $W_\upsilon(\theta)$ are synthesized and the corresponding 2D normalized intensity distributions along the transverse planes shifted from the focal plane have been calculated analytically and are plotted for orders, $\upsilon = 0, 1, 2, \ldots, 7$ and are presented in the following section. The results are validated with the known result of Airy pupil.

4.5 Illustrative Results with Discussion

4.5.1 Intensity Distributions in the Far-Field Region for Zero Order Azimuthal Walsh Filters

Intensity distribution at the far-field plane due to azimuthal Walsh filters of order zero placed at the exit pupil plane of a rotationally symmetric imaging system corresponds to that due to the Airy pupil as shown in Fig. 2.12. The intensity at the origin of the far-field plane is maximum and the intensity distribution along the axis is symmetric around the far-field plane.

The variation of intensity of the centre of the diffraction pattern for zero order azimuthal Walsh filter placed on the exit pupil plane, at different longitudinally shifted image planes along the far-field region, has been calculated and plotted against the defocus aberration term, W_{20} and is shown in Fig. 4.7.

The values of W_{20} considered here, are ranging from −5 to +5 with successive steps of 0.5, where the intensity is seen to be completely exhausted beyond $W_{20} = \pm 5$. The graph shows evidence of defocusing of Airy pupil or the well-known phenomenon of Fresnel diffraction at the vicinity of the focal plane.

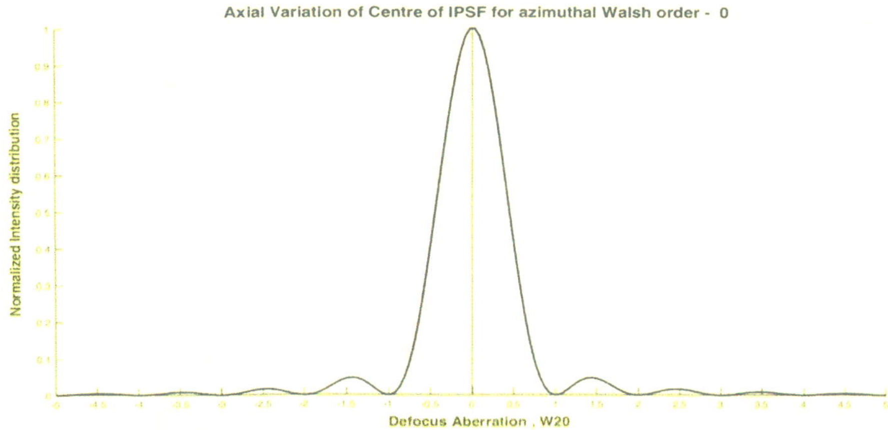

Fig. 4.7 Axial variation of centre of the diffraction patterns for zero order azimuthal Walsh function, at shifted transverse planes, with defocusing aberration term, W_{20} ranging from −5 to +5

4.5.2 Intensity Distributions in the Far-Field Region for Higher Order Azimuthal Walsh Filters

With computational results for zero order azimuthal Walsh filters exactly matching with the existing results for Airy pupil, we have proceeded for computation of intensity point spread functions for higher order azimuthal Walsh filters $W_\nu(\theta)$, for order $\nu = 1, 2, \ldots, 7$, taking into consideration of defocus aberration term W_{20} varying from $-5 \leq W_{20} \leq 0$ and $0 \leq W_{20} \leq +5$ with successive steps of 0.5. Tables 4.1, 4.2, 4.3, 4.4, 4.5, 4.6 and 4.7 represent the polar plots of 2D IPSF, contour IPSF and snapshots of 3D IPSFs generated analytically from 2D intensity distributions due to azimuthal Walsh filters $W_\nu(\theta)$ for orders $\nu = 1, 2, 3, 4, 5, 6$ and 7 respectively, for defocus aberration term W_{20} varying from $0 \leq W_{20} \leq +5$, i.e., moving away from the far-field plane in the (+)ve direction.

The results presented above for higher order azimuthal Walsh filters for $0 \leq W_{20} \leq +5$ are found to be exactly similar to those calculated for variation of defocus aberration term W_{20} from $-5 \leq W_{20} \leq 0$, i.e., moving away from the far-field plane in the (−)ve direction. An inspection to the tables above reveals that the axial intensity on the far-field plane is zero for all orders and the intensity distribution along the axis is asymmetric. It is also observed that the intensity distributions along transverse planes spread out with the increase in the value of W_{20}.

Table 4.1 PSFs due to $W_1(\theta)$ on transverse planes axially shifted in the (+) ve direction away from the far-field plane, i.e. for $0 \leq W_{20} \leq +5$

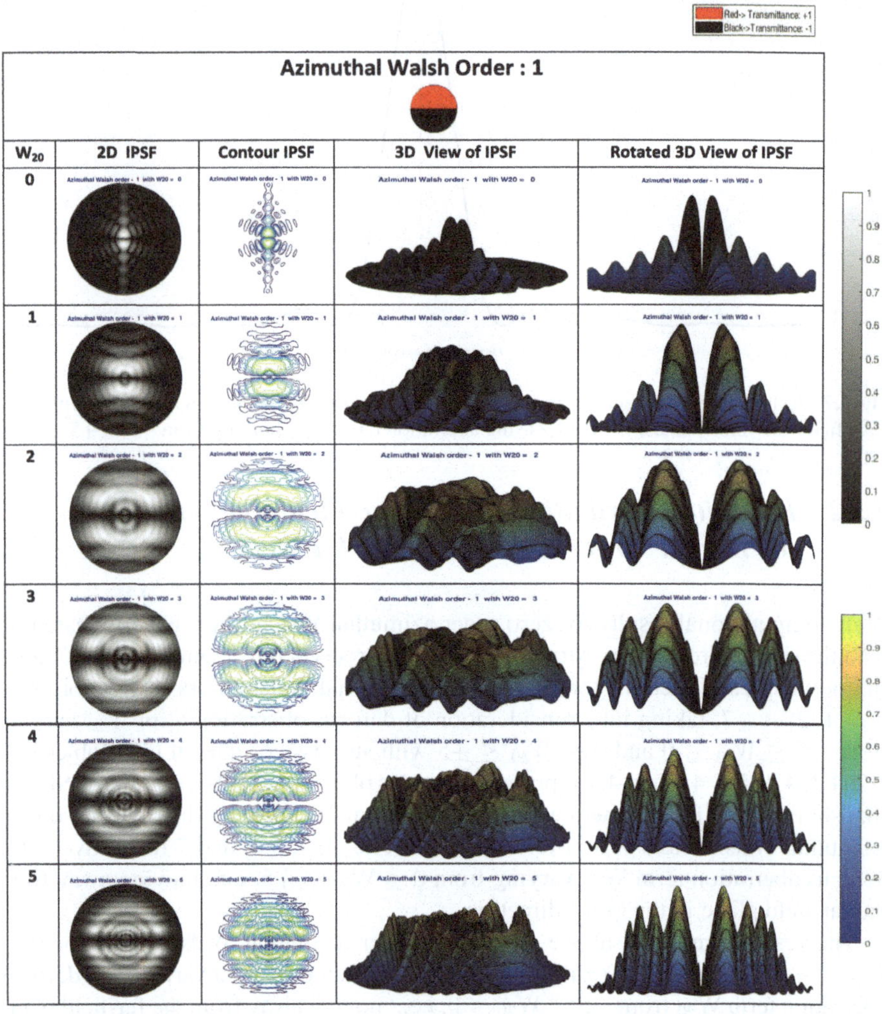

4.6 Intensity Distributions on Transverse Planes Very Near to the Focus with Azimuthal Walsh Filters

The light distributions in the immediate vicinity of the focal plane of the rotationally symmetric imaging system under consideration, with azimuthal Walsh filters $W_{\upsilon}(\theta)$ of orders, $\nu = 1, 2, 3, 4, \ldots, 7$ placed on the exit pupil plane (refer to Fig. 4.1) are studied to observe the more accurate effect of wavefront aberration on diffraction pattern and are presented below.

Table 4.2 PSFs due to $W_2(\theta)$ on transverse planes axially shifted in the (+) ve direction away from the far-field plane, i.e. for $0 \leq W_{20} \leq +5$

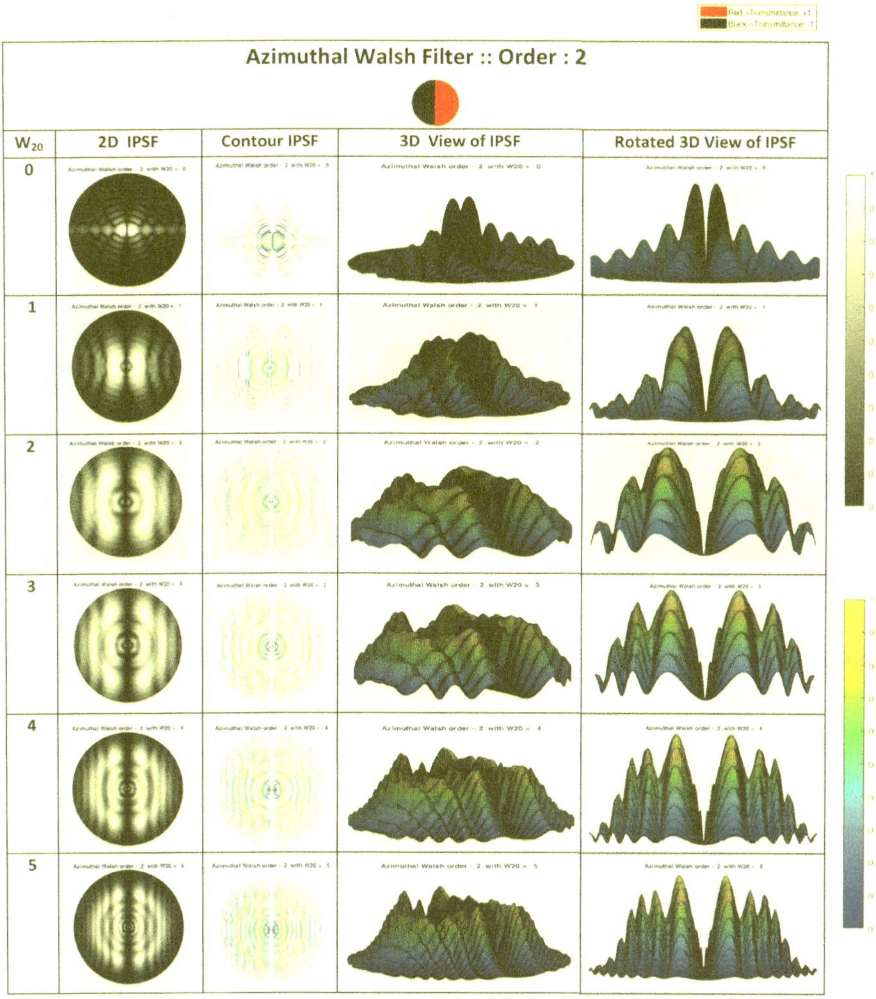

Near field intensity distribution in the immediate vicinity of the focal plane for $W_\nu(\theta)$, for order $\nu = 1$, placed on the exit pupil plane has been calculated and presented in Tables 4.8 and 4.9 for $0 \leq W_{20} \leq +1.25$ and $+1.5 \leq W_{20} \leq +2$ respectively in successive smaller steps of 0.25 to capture the minute variation of intensity distribution due to corresponding change in defocus Aberration, W_{20}.

Tables 4.10 and 4.11 present polar plots of 2D IPSF, contour IPSF and snapshots of 3D IPSF and rotated view at the near focus transverse planes, showing zero intensity at the centre of the intensity distribution for second order azimuthal Walsh filter, $W_2(\theta)$, for $0 \leq W_{20} \leq +1.25$, and $+1.5 \leq W_{20} \leq +2$, with successive smaller steps of 0.25.

Table 4.3 PSFs due to $W_3(\theta)$ on transverse planes axially shifted in the (+) ve direction away from the far-field plane, i.e. for $0 \leq W_{20} \leq +5$

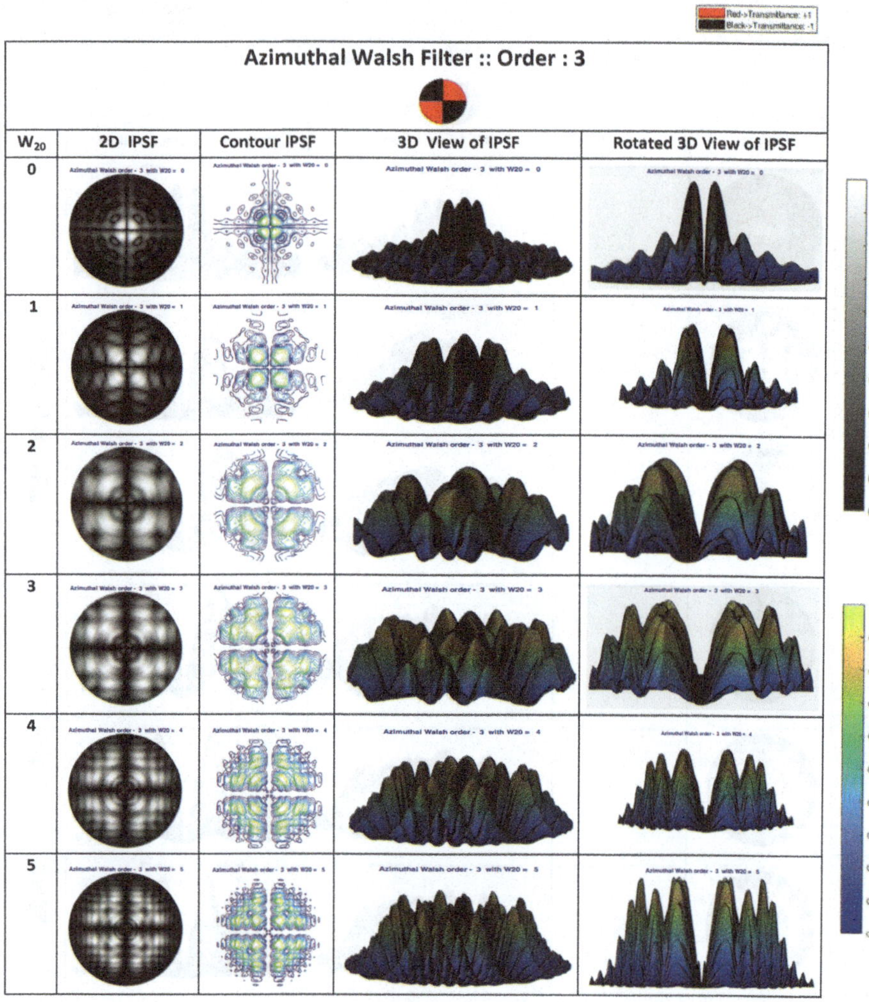

As per Tables 4.12 and 4.13, the axial intensity on the far-field plane for $W_3(\theta)$ is zero and the intensity distribution along the axis is asymmetric around the far-field plane as in $W_2(\theta)$ and $W_1(\theta)$. The intensity distribution along the axis is same for both $+\Delta Z$ and $-\Delta Z$, where $+\Delta Z$ represents axial shift of the transverse plane along the (+)ve direction and $-\Delta Z$ is the corresponding shift along the (−)ve direction of the far-field plane, considering the position of focal plane at the centre of the coordinate system.

Table 4.4 PSFs due to $W_4(\theta)$ on transverse planes axially shifted in the (+) ve direction away from the far-field plane, i.e. for $0 \leq W_{20} \leq +5$

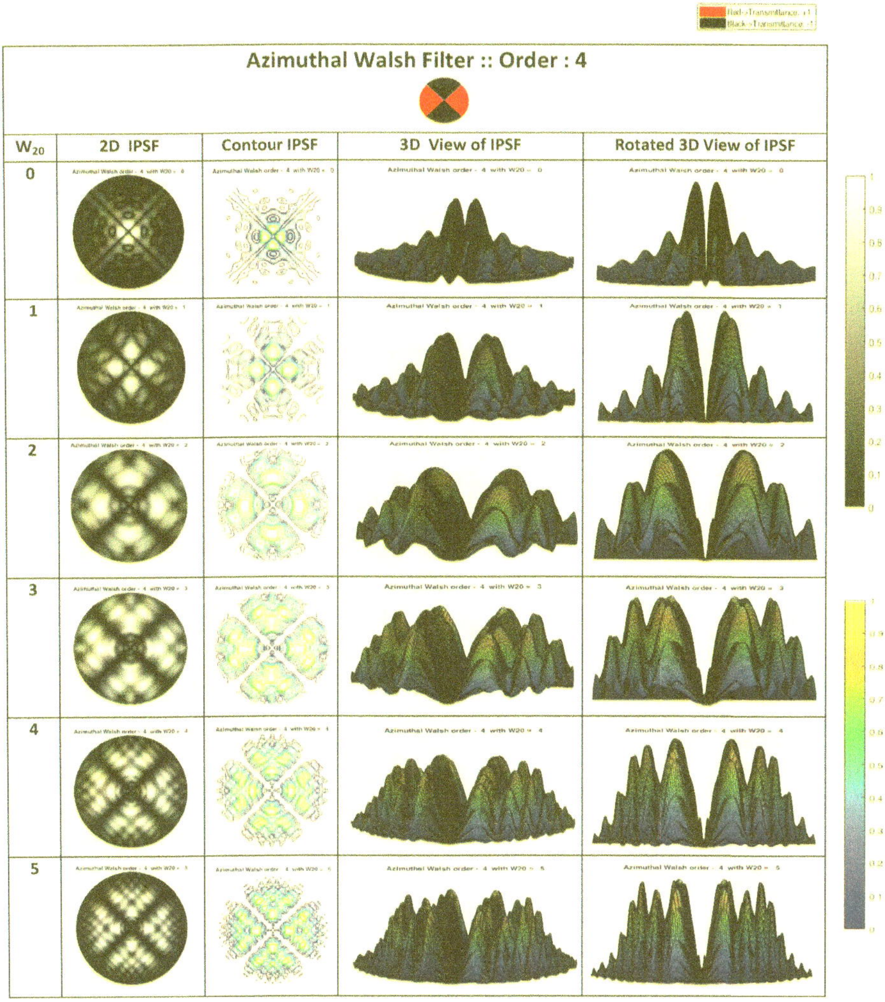

Tables 4.14 and 4.15 present the near field intensity distributions for fourth order azimuthal Walsh filter $W_4(\theta)$, placed on the exit pupil plane of the rotationally symmetric imaging system under consideration, has been calculated and presented for different values of defocus aberration term W_{20} varying from $0 \leq W_{20} \leq +1.25$ and $1.5 \leq W_{20} \leq +2$ respectively.

Table 4.5 PSFs due to $W_5(\theta)$ on transverse planes axially shifted in the (+) ve direction away from the far-field plane, i.e. for $0 \le W_{20} \le +5$

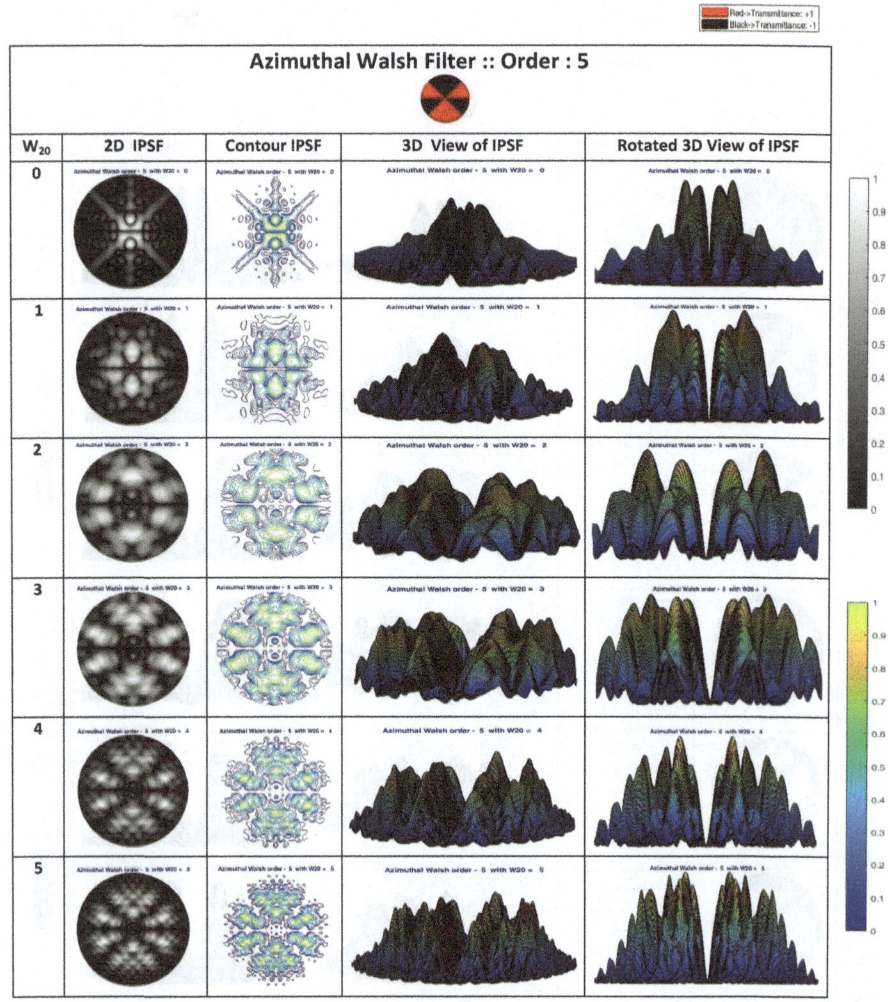

Tables 4.16 and 4.17 present polar plots of 2D IPSF, contour IPSF, 3D IPSF and rotated view of 3D IPSF at the near focus transverse planes, showing zero intensity at the centre of the intensity distribution for fifth order azimuthal Walsh filter $W_5(\theta)$, for variation of defocus aberration term W_{20} as $0 \le W_{20} \le +1.25$ and $1.5 \le W_{20} \le +2$, respectively, with successive steps of 0.25.

Table 4.6 PSFs due to $W_6(\theta)$ on transverse planes axially shifted in the (+) ve direction away from the far-field plane, i.e. for $0 \leq W_{20} \leq +5$

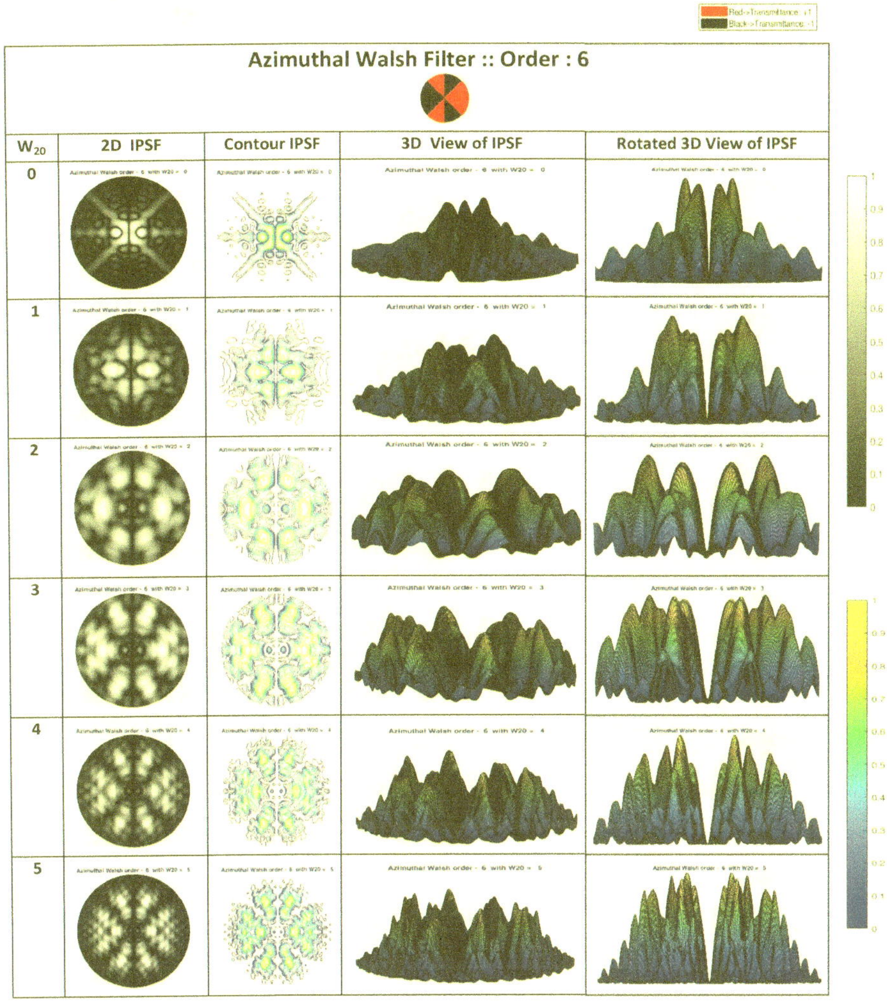

Tables 4.18 and 4.19 present polar plots of 2D IPSF, contour IPSF, 3D IPSF and rotated view of 3D IPSF at the near focus transverse planes, showing zero intensity at the centre of the intensity distribution for sixth order azimuthal Walsh filter $W_6(\theta)$, for variation of defocus aberration term W_{20} as $0 \leq W_{20} \leq +1$ and $1.25 \leq W_{20} \leq +2$, respectively, with successive steps of 0.25.

Table 4.7 PSFs due to $W_7(\theta)$ on transverse planes axially shifted in the (+) ve direction away from the far-field plane, i.e. for $0 \leq W_{20} \leq +5$

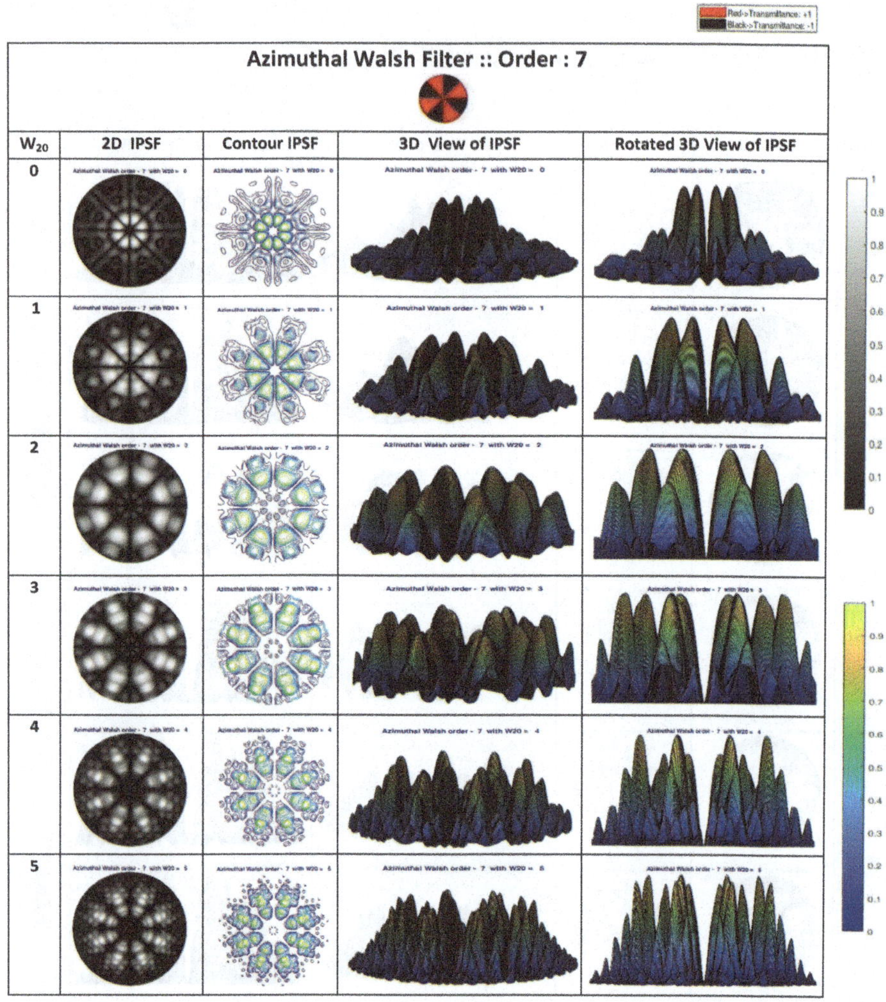

Tables 4.20 and 4.21 present polar plots of 2D IPSF, contour IPSF, 3D IPSF and rotated view of 3D IPSF at the near focus transverse planes, for $W_7(\theta)$, for variation of defocus aberration term W_{20} as $0 \leq W_{20} \leq +1.25$ and $+1.5 \leq W_{20} \leq +2$ respectively, with successive steps of $+0.25$.

Table 4.8 PSFs due to $W_1(\theta)$ on transverse planes axially shifted in the (+) ve direction away from the far-field plane, i.e. for $0 \leq W_{20} \leq +1.25$

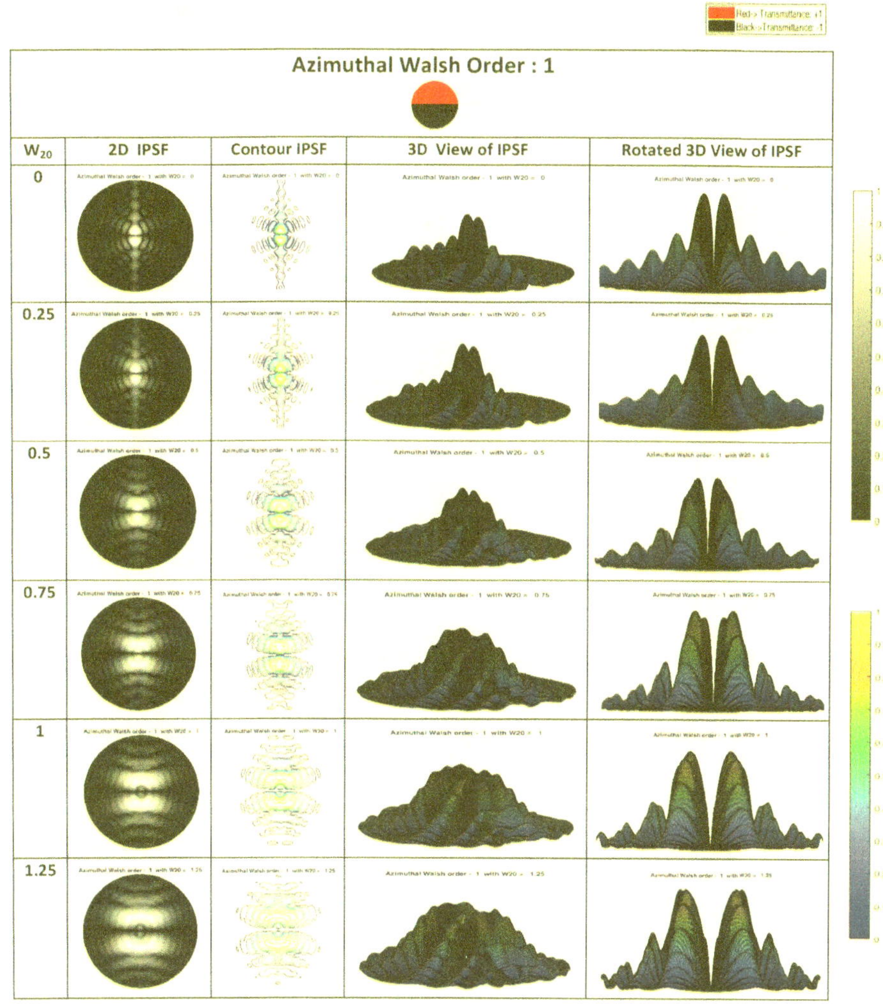

It is calculated that the change in diffraction pattern due to change in the values of W_{20} from $-2 \leq W_{20} \leq 0$ is exactly same as that due to change in the values of W_{20} as $0 \leq W_{20} \leq +2$. This means the diffraction pattern on the near focal transverse planes changes in the same way as one proceed away from the far-field plane along the (+) ve or (−) ve direction.

Table 4.9 PSFs due to $W_1(\theta)$ on transverse planes axially shifted in the (+) ve direction away from the far-field plane, i.e. for $0 \leq W_{20} \leq +2$

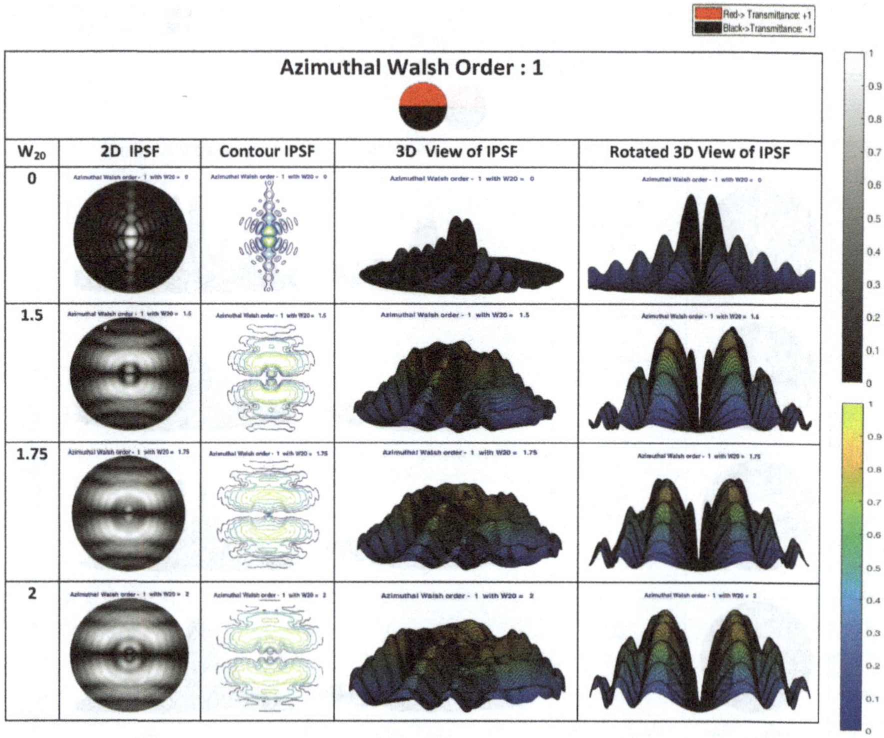

So, it is apparent that light can be efficiently sculpted around the far-field region of the rotationally symmetric imaging system (refer to Fig. 4.1) with azimuthal Walsh filters at the exit pupil plane. This can be efficiently used to form gradient force trap required for conventional as well as plasmonic optical tweezers [2–5] for manipulating objects of size comparable to a single atom to 100 μm without mechanical contact.

Table 4.10 PSFs due to $W_2(\theta)$ on transverse planes axially shifted in the (+) ve direction away from the far-field plane, i.e. for $0 \le W_{20} \le 1.25$

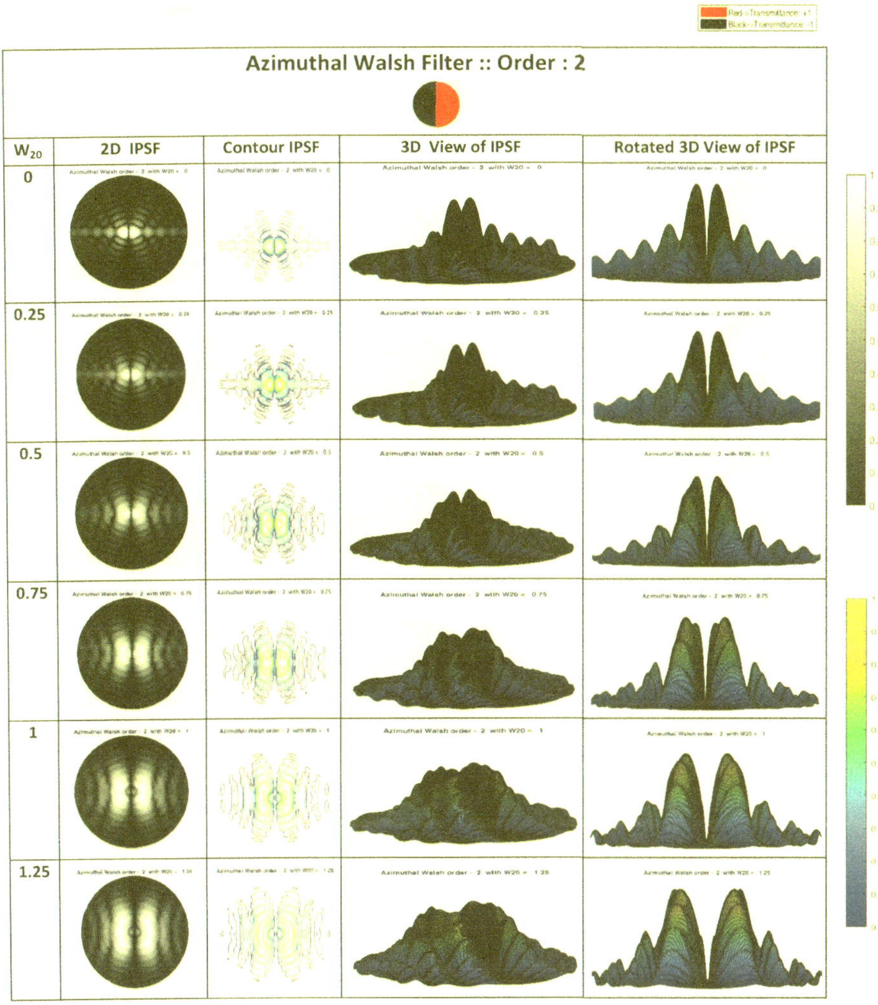

Table 4.11 PSFs due to $W_2(\theta)$ on transverse planes axially shifted in the (+) ve direction away from the far-field plane, i.e. for $0 \leq W_{20} \leq +2$

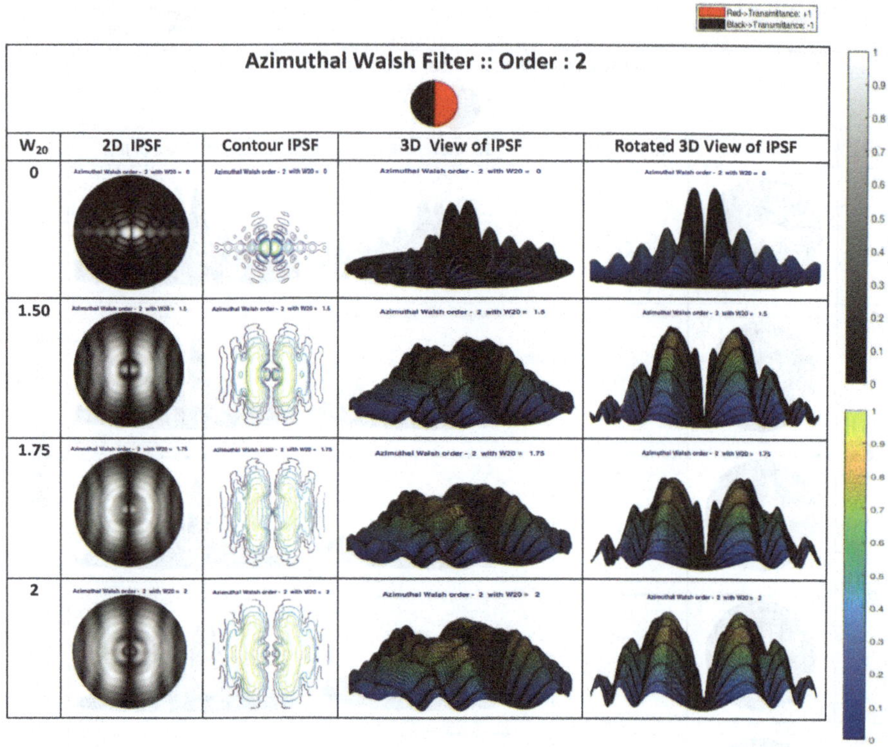

Table 4.12 PSFs due to $W_3(\theta)$ on transverse planes axially shifted in the (+) ve direction away from the far-field plane, i.e. for $0 \leq W_{20} \leq +1.25$

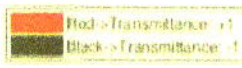

Azimuthal Walsh Filter :: Order : 3

W_{20}	2D IPSF	Contour IPSF	3D View of IPSF	Rotated 3D View of IPSF
0	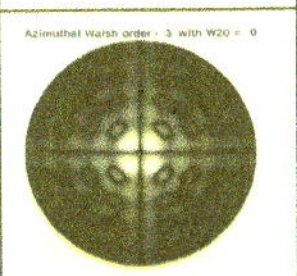	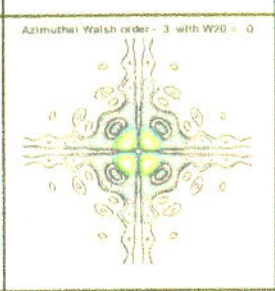		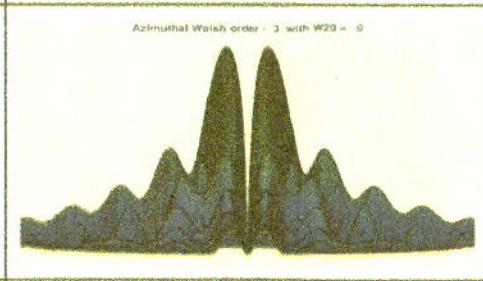
0.25	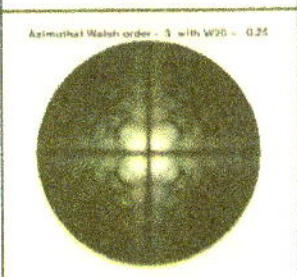		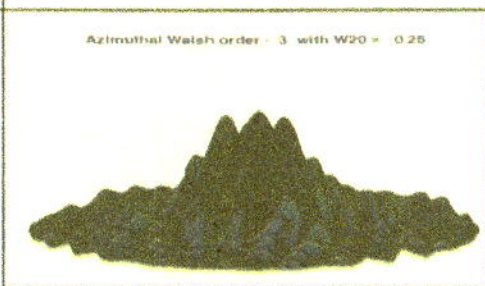	
0.5	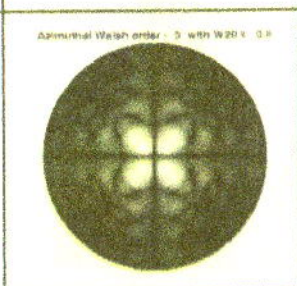	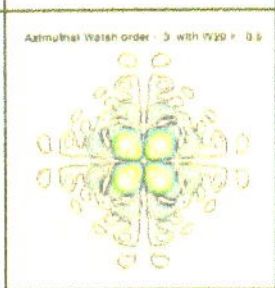	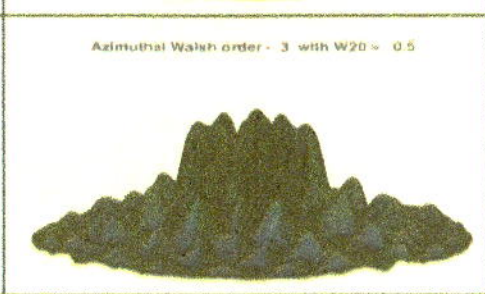	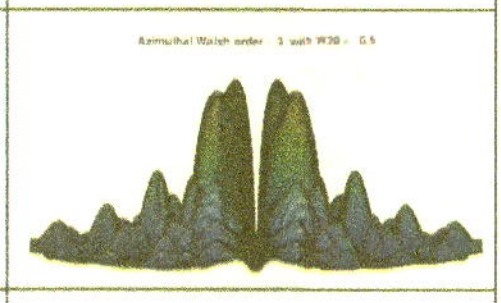
0.75	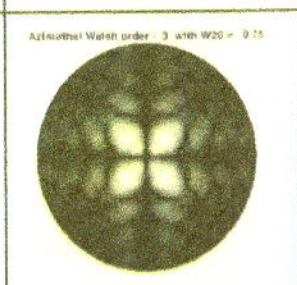	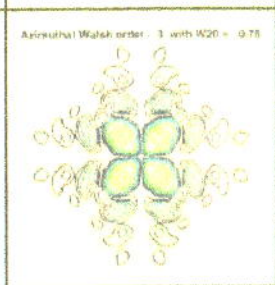	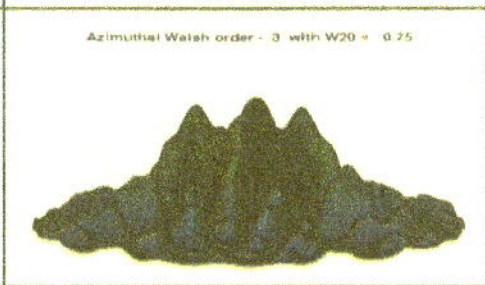	
1	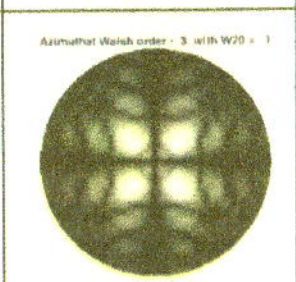	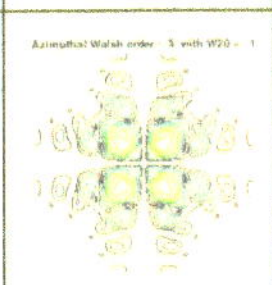	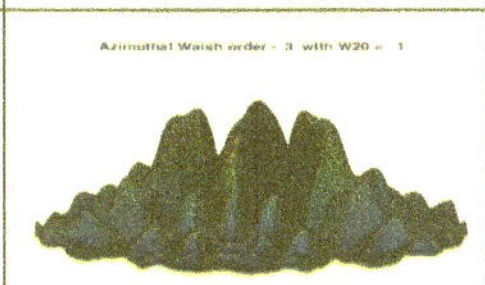	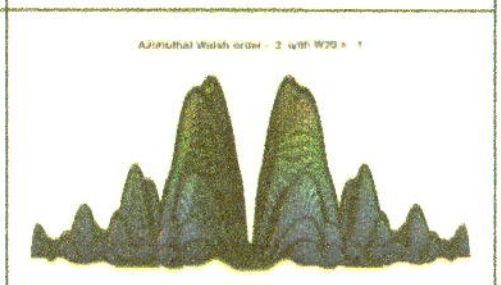
1.25	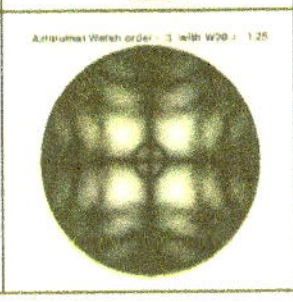	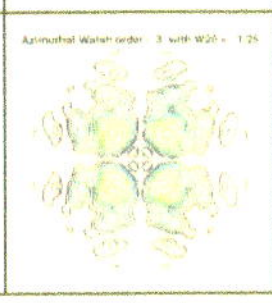	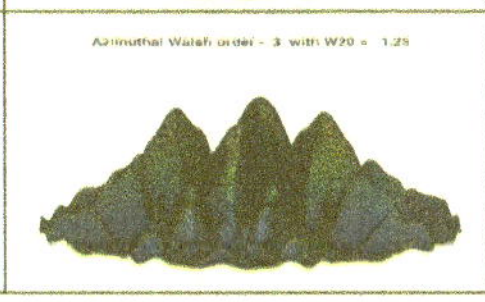	

Table 4.13 A PSFs due to $W_3(\theta)$ on transverse planes axially shifted in the (+) ve direction away from the far-field plane, i.e. for $0 \leq W_{20} \leq +2$

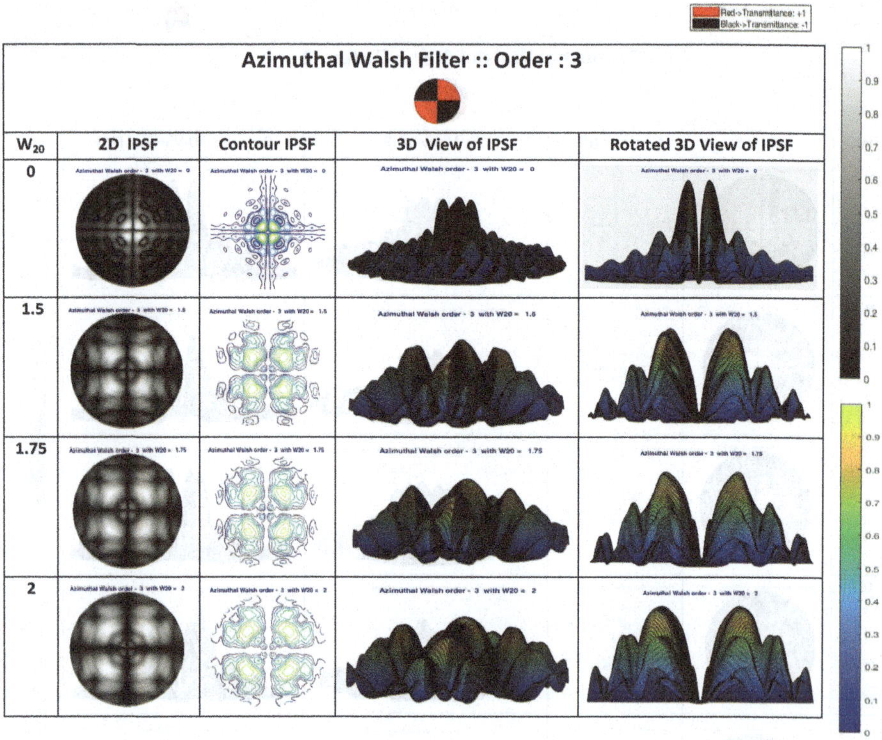

Table 4.14 PSFs due to $W_4(\theta)$ on transverse planes axially shifted in the (+) ve direction away from the far-field plane, i.e. for $0 \le W_{20} \le +1.25$

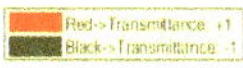

Azimuthal Walsh Filter :: Order : 4

W_{20}	2D IPSF	Contour IPSF	3D View of IPSF	Rotated 3D View of IPSF
0	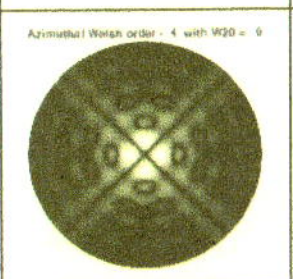	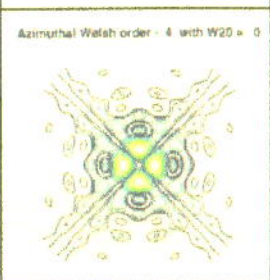	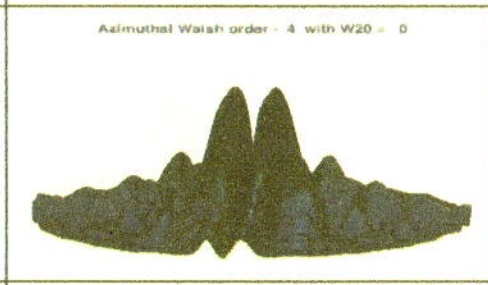	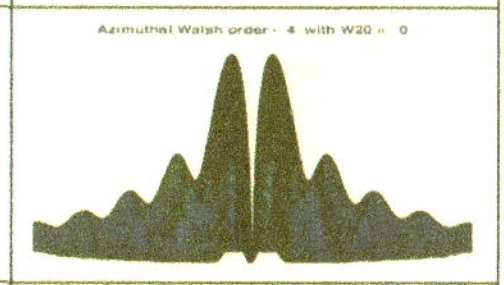
0.25		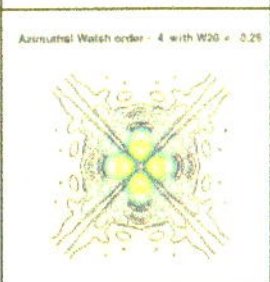	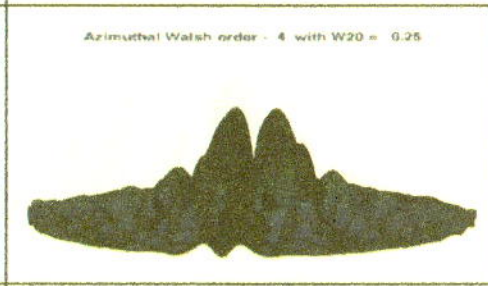	
0.50	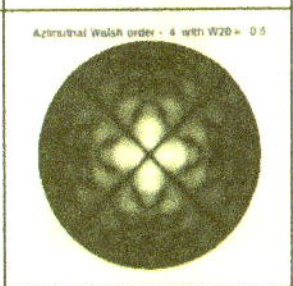	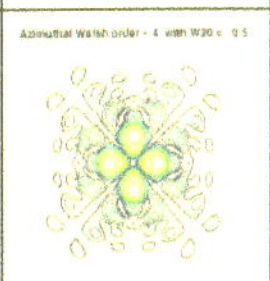	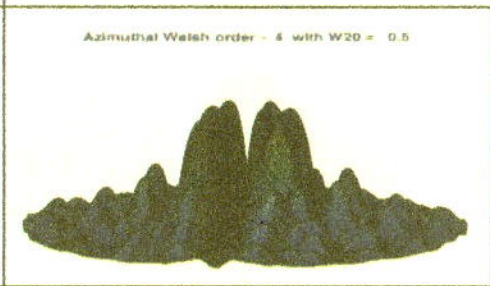	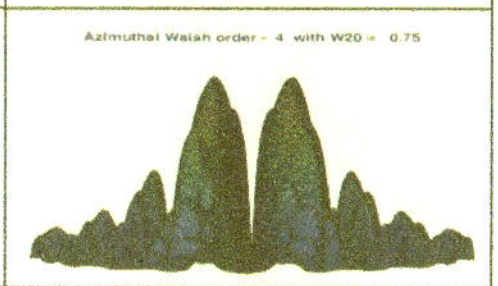
0.75	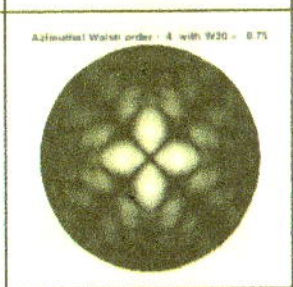	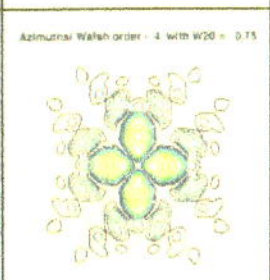	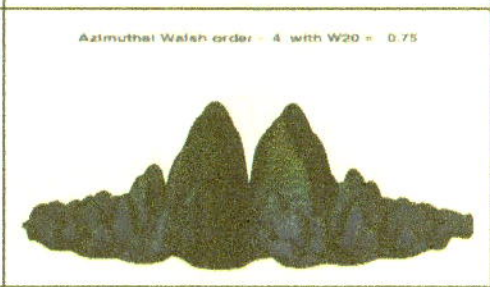	
1	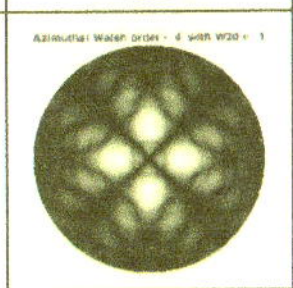	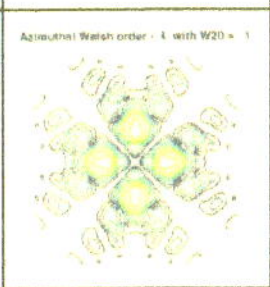	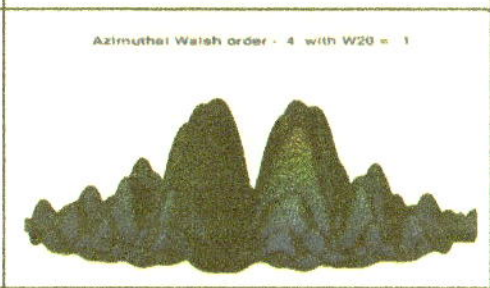	
1.25	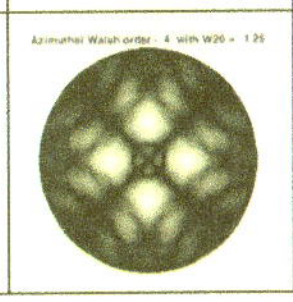	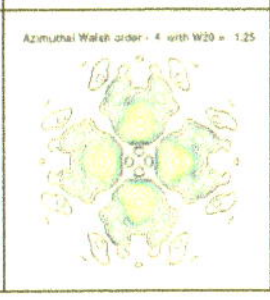	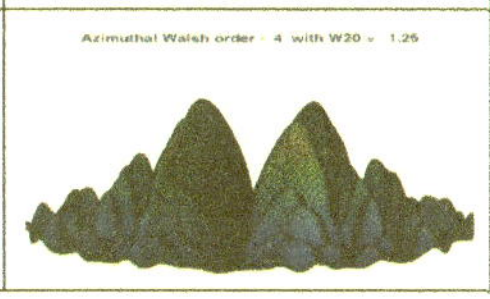	

Table 4.15 PSFs due to $W_4(\theta)$ on transverse planes axially shifted in the (+) ve direction away from the far-field plane, i.e. for $0 \leq W_{20} \leq +2$

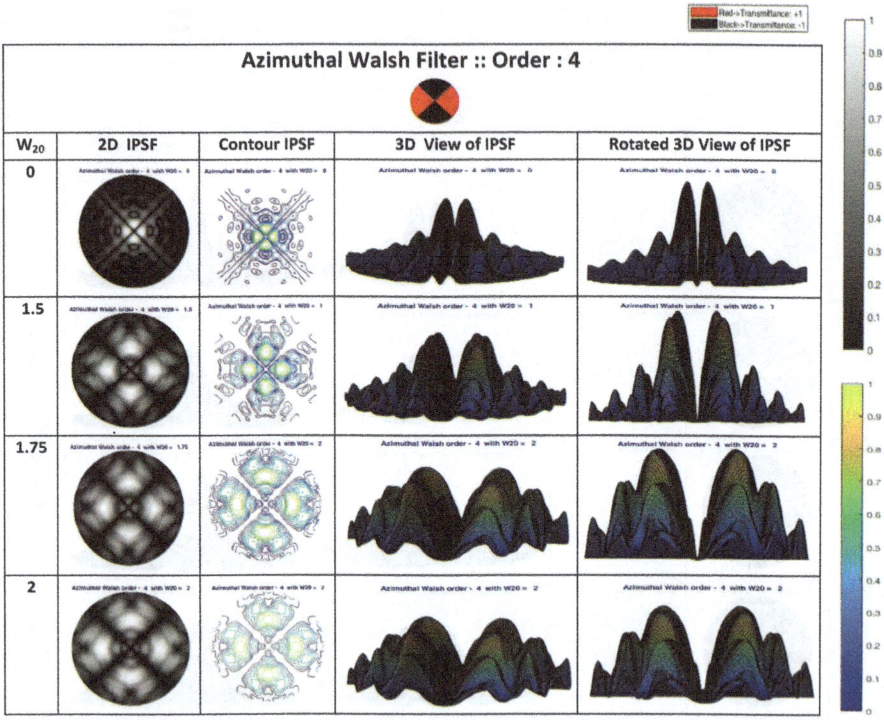

Table 4.16 PSFs due to $W_5(\theta)$ on transverse planes axially shifted in the (+) ve direction away from the far-field plane, i.e. for $0 \le W_{20} \le +1.25$

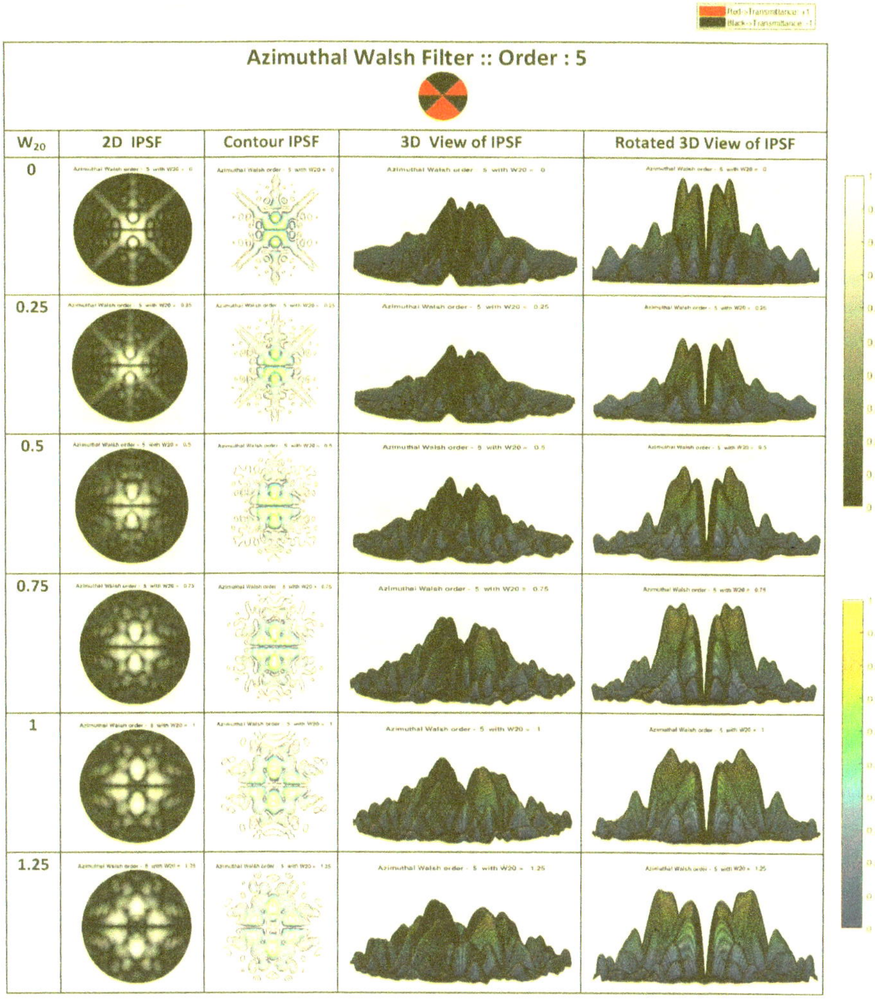

Table 4.17 PSFs due to $W_5(\theta)$ on transverse planes axially shifted in the (+) ve direction away from the far-field plane, i.e. for $0 \leq W_{20} \leq +2$

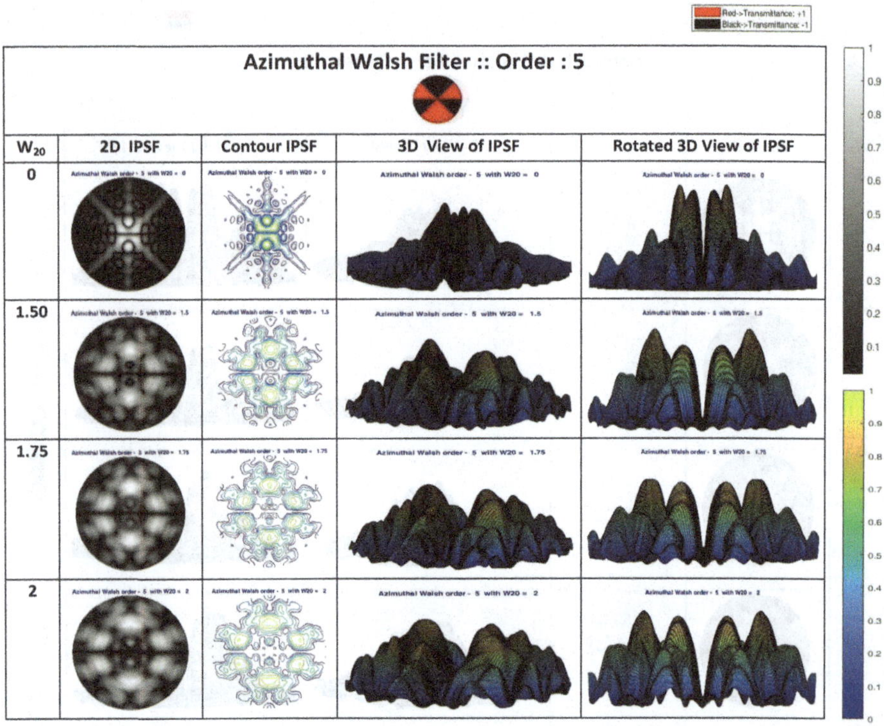

Table 4.18 PSFs due to $W_6(\theta)$ on transverse planes axially shifted in the (+) ve direction away from the far-field plane, i.e. for $0 \leq W_{20} \leq +1.25$

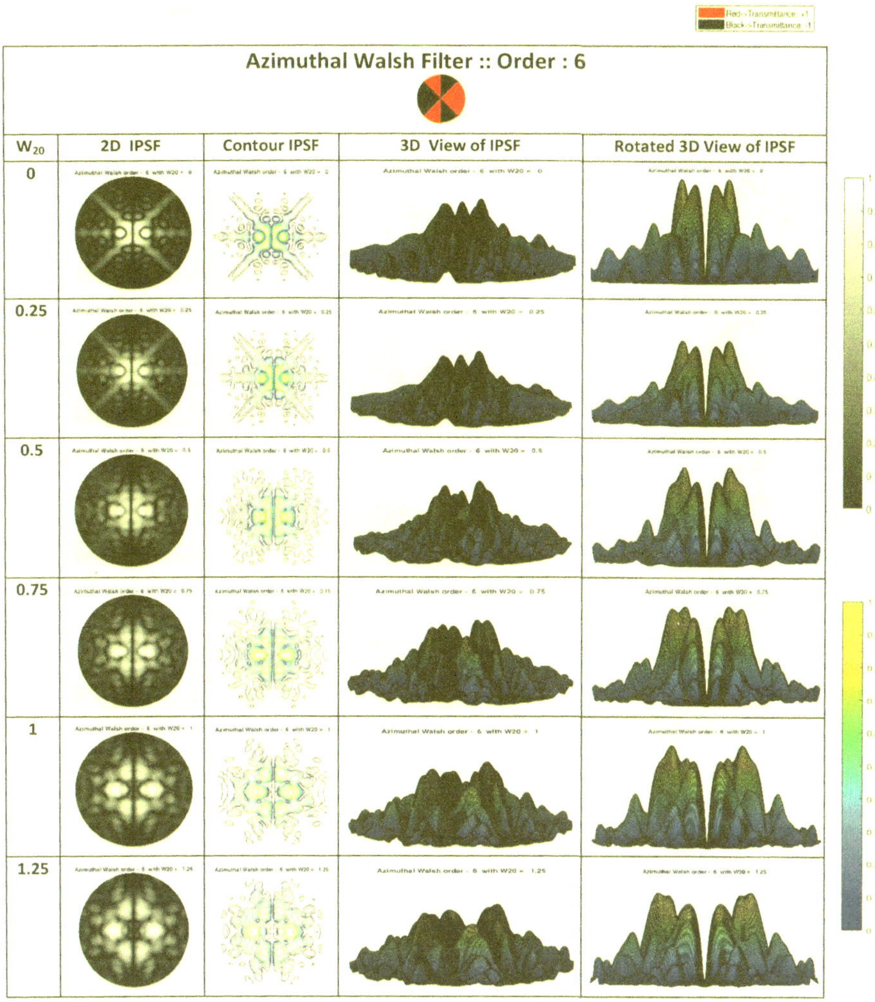

Table 4.19 PSFs due to $W_6(\theta)$ on transverse planes axially shifted in the (+) ve direction away from the far-field plane, i.e. for $0 \leq W_{20} \leq +2$

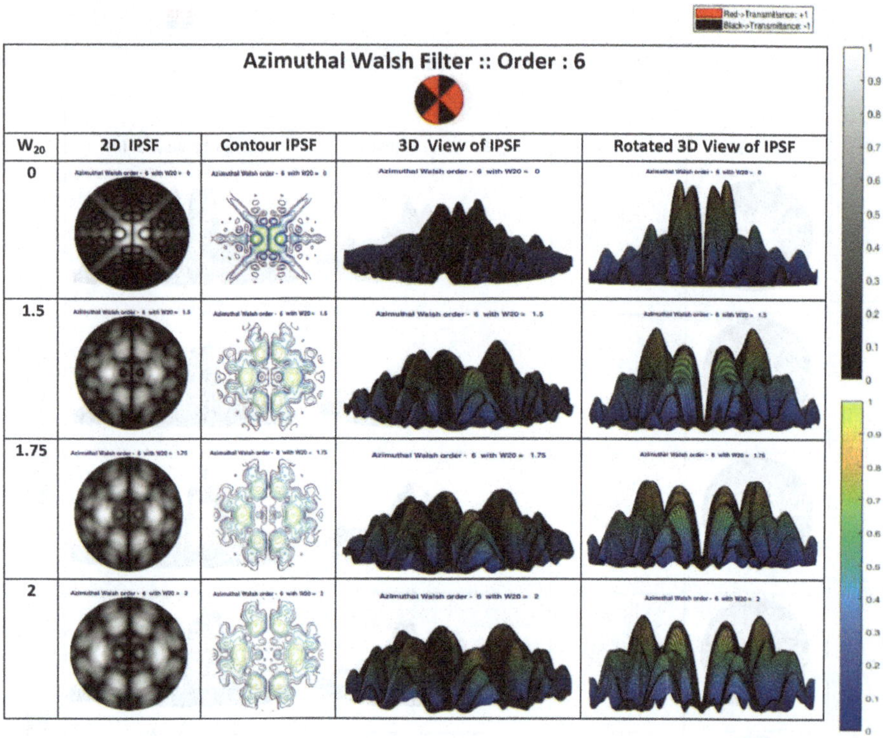

Table 4.20 PSFs due to $W_7(\theta)$ on transverse planes axially shifted in the (+) ve direction away from the far-field plane, i.e. for $0 \leq W_{20} \leq +5$

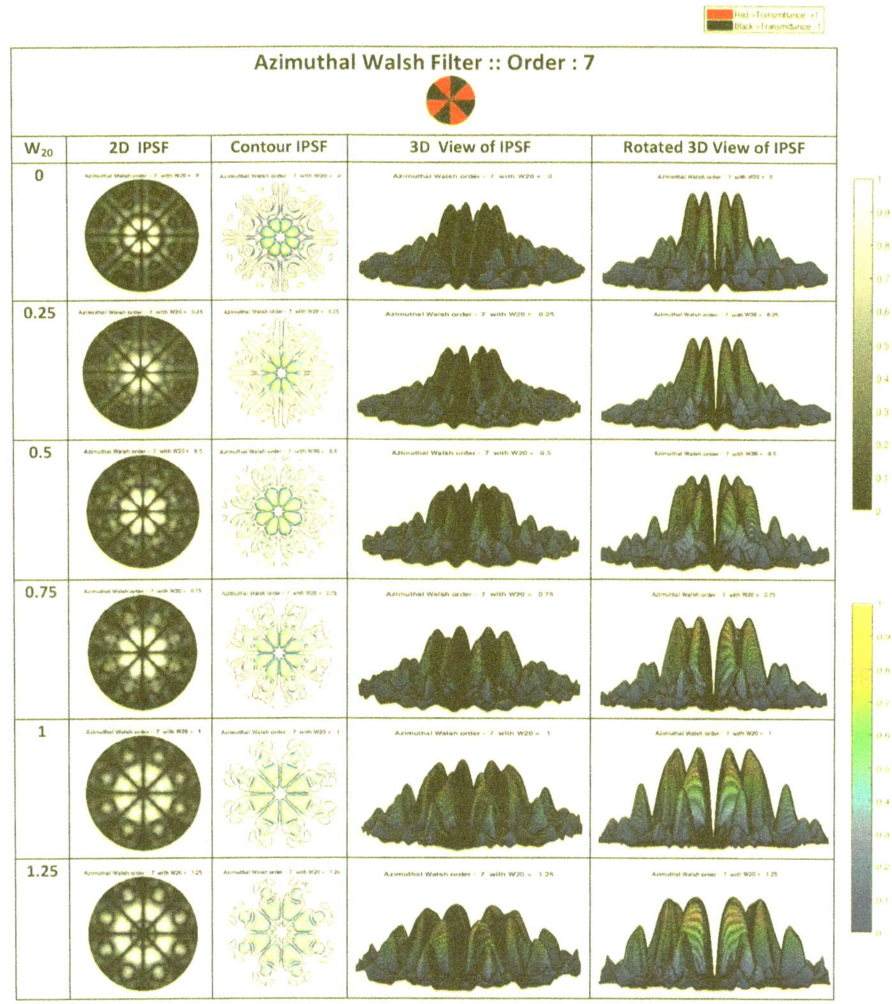

Table 4.21 PSFs due to $W_7(\theta)$ on transverse planes axially shifted in the (+) ve direction away from the far-field plane, i.e. for $0 \leq W_{20} \leq +2$

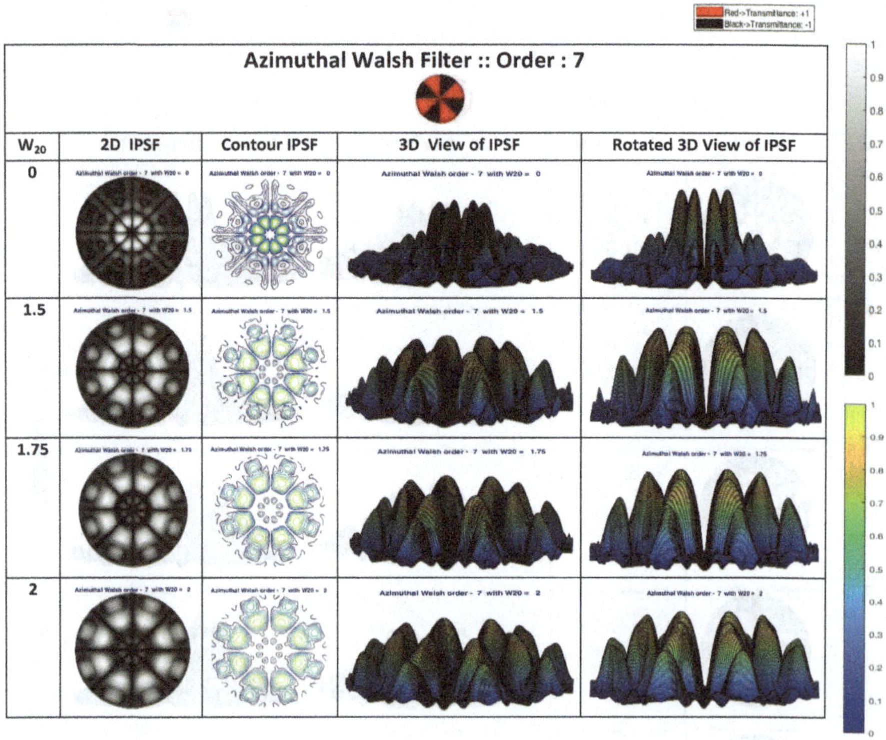

References

1. L. Hazra, P. Mukherjee, Self-similarity in Walsh Functions and in the Farfield Diffraction Patterns of Radial Walsh Filters, SpringerBriefs in Applied Sciences and Technology (2018). https://doi.org/10.1007/978-981-10-2809-0_5
2. P.N. Lebedev, Experimental examination of light pressure. Ann. Phys. **6**, 433–458 (1901)
3. A. Ashkin, Acceleration and trapping of particles by radiation pressure. Phy. Rev. Lett. **24**(4), 156–159 (1970)
4. D. Grass, J. Fesel, S. Hofer, N. Kiesel, M. Aspelmeyer, Optical trapping and control of nanoparticles inside evacuated hollow core photonic crystal fibres. Appl. Phys. Lett. **108**, 221103 (2016)
5. S. Tatsuya, T. Yasuyuki, Plasmonic optical tweezers toward molecular manipulation: tailoring plasmonic nanostructure, light source, and resonant trapping. J. Phys. Chem. Lett. **5**, 2957–2967 (2014)

Chapter 5
Self-Similarity in Transverse Intensity Distributions in the Far-Field Region of Self-Similar Azimuthal Walsh Filters

5.1 Introduction

Self-similarity in transverse intensity distributions on the far-field plane has been observed within self-similar members of distinct groups and subgroups of azimuthal Walsh filters [1, 2] and it has been illustrated and discussed extensively in Chap. 3. Self-similarity has also been observed in the transverse intensity distributions at shifted image planes along the far-field region by the members of self-similar Groups and Sub-Groups of azimuthal Walsh filters. Azimuthal Walsh filters of self-similar group members are placed on the exit pupil of a rotationally symmetric imaging system (Fig. 4.1) and the normalized intensity distributions on longitudinally shifted image planes from the paraxial focal plane are computed. The image planes are shifted longitudinally from the paraxial focal plane or far-field plane by an actual amount ΔZ, which causes defocus or 'Depth of Focus' aberration represented by the term W_{20}. W_{20} is a measure of asphericity of the image forming wavefront as shown in Fig. 4.2. Transverse intensity distributions at shifted image planes, around the far-field plane, for self-similar members of azimuthal Walsh filters $W_{\nu}(\theta)$, for orders $\nu = 1, 2, 3, \ldots$, have been calculated and presented in this chapter group and sub-group-wise and for different values of defocus aberration term $W_{20,}$ within ranges $-5 \leq W_{20} \leq 0$ and $0 \leq W_{20} \leq +5$ with a regular interval of ± 1.

I. Bhattacharya and L. Hazra, *Azimuthal Walsh Filters*, Progress in Optical Science and Photonics 10,
https://doi.org/10.1007/978-981-15-6099-6_5

5.2 Self-Similarity in Transverse Intensity Distributions in the Far-Field Region for Group I Self-Similar Members of Azimuthal Walsh Filters

Azimuthal Walsh filters of Group I consisting of orders 1, 3, 7, … observed to produce self-similar transverse intensity distributions around the far-field region, when calculated for any particular azimuth. Transverse intensity distributions around the far-field plane due to Group I self-similar members, for a particular azimuth, $\zeta = 0$ taking into consideration of defocus aberration term W_{20} within the ranges of $0 \leq W_{20} \leq +5$ and $-5 \leq W_{20} \leq 0$ have been illustrated in this section. The axial intensity on the shifted image planes is observed to be zero. In absence of defocus aberration term, i.e., for $W_{20} = 0$, at the far-field plane, the transverse intensity distributions at shifted image planes, due to self-similar members of Group I azimuthal Walsh filters, for orders $\nu = 1, 3$ and 7 are illustrated in Fig. 5.1a–c respectively, for azimuth $\zeta = 0$.

In absence of defocus, $W_{20} = 0$ i.e., on the far-field plane, the normalized intensity distributions exhibited by Group I self-similar members of azimuthal Walsh filters show self-similar pattern with central minima. The central minima widen with the increase of order number, ν. The intensity of the side lobes increases with the increase of ν of the members of Group I as observed for azimuth, $\zeta = 0$.

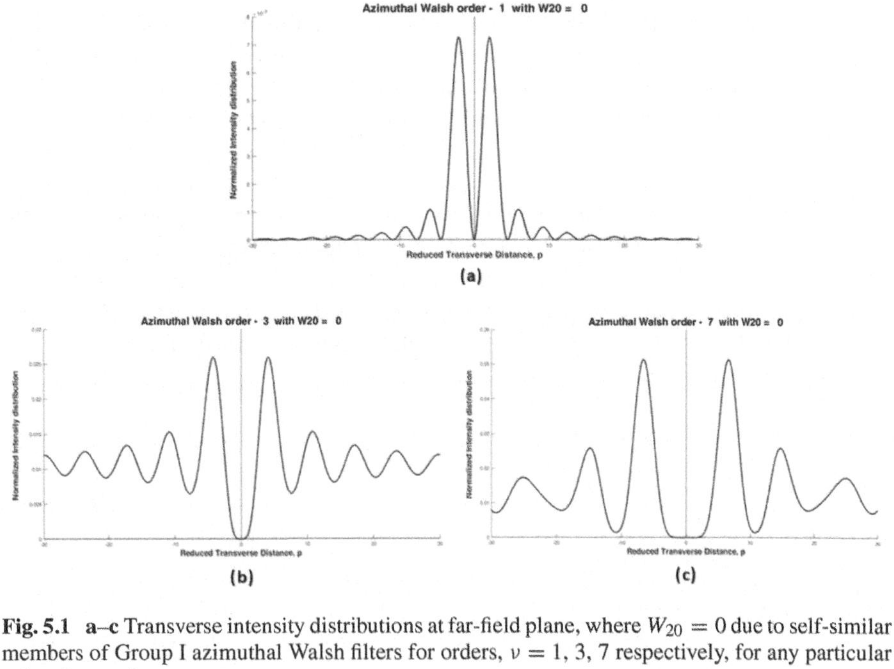

Fig. 5.1 **a–c** Transverse intensity distributions at far-field plane, where $W_{20} = 0$ due to self-similar members of Group I azimuthal Walsh filters for orders, $\nu = 1, 3, 7$ respectively, for any particular azimuth, $\zeta = 0$

5.2.1 *Study of Self-Similarity on Transverse Image Planes Shifted Towards Right or (+)ve Side of Far-Field Plane*

Study of self-similarity in the transverse intensity distributions on the longitudinally shifted image planes from the far-field plane by an amount $+\Delta Z$, where $+\Delta Z$ represents longitudinal shift of the image plane from the far-field plane along (+)ve direction, has been presented in this section. The corresponding defocus or 'Depth of Focus' aberration term W_{20} in this region has been considered to be within the range $0 \leq W_{20} \leq +5$ with successive steps of 1.

Case 1: For $W_{20} = 1$

For the value of defocus aberration term $W_{20} = 1$, the transverse intensity distributions due to Group I self-similar members of azimuthal Walsh filters for orders $\nu = 1, 3$ and 7 are illustrated in Fig. 5.2a–c respectively for azimuth, $\zeta = 0$.

Case 2: For $W_{20} = 2$

For the value of defocus aberration term $W_{20} = 2$, the transverse intensity distributions due to self-similar members of Group I azimuthal Walsh filters for orders $\nu = 1, 3$ and 7 are illustrated in Fig. 5.3a–c respectively, for azimuth $\zeta = 0$.

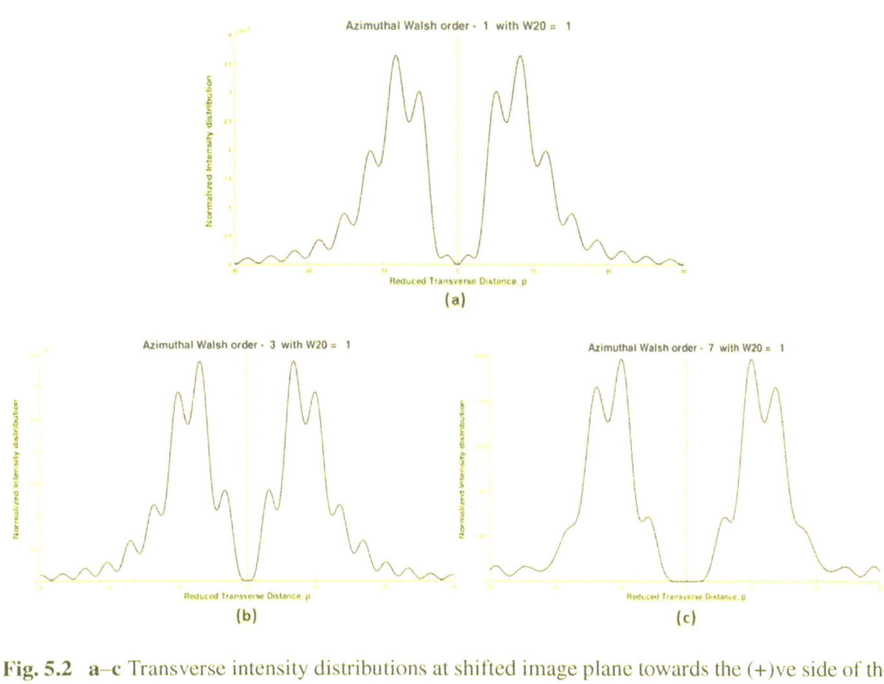

Fig. 5.2 **a–c** Transverse intensity distributions at shifted image plane towards the (+)ve side of the far-field plane with $W_{20} = 1$, due to self-similar members of Group I azimuthal Walsh filters, for orders, $\nu = 1, 3$ and 7 respectively, for any particular azimuth, $\zeta = 0$

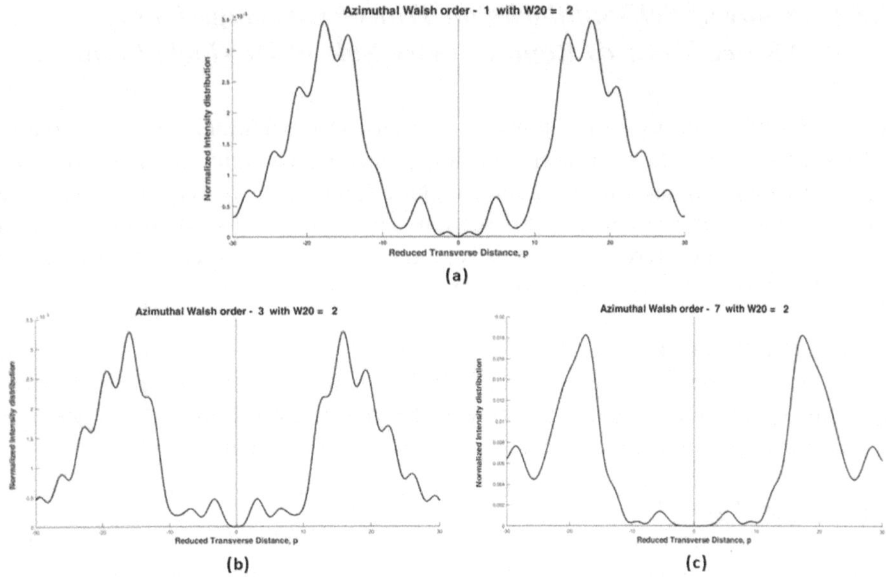

Fig. 5.3 a–c Transverse intensity distributions at shifted image plane with $W_{20} = 2$, due to self-similar members of Group I azimuthal Walsh filters for orders, $\nu = 1, 3$ and 7 respectively, for any particular azimuth $\zeta = 0$

Case 3: For $W_{20} = 3$

For the value of defocus aberration term $W_{20} = 3$, the transverse intensity distributions due to self-similar members of Group I azimuthal Walsh filters for orders, $\nu = 1, 3$ and 7 are illustrated in Fig. 5.4a–c respectively, for azimuth $\zeta = 0$.

Case 4: For $W_{20} = 4$

For a defocus aberration term $W_{20} = 4$, the transverse intensity distributions due to self-similar members of Group I azimuthal Walsh filters for orders $\nu = 1, 3$ and 7 are illustrated in Fig. 5.5a–c respectively, for azimuth $\zeta = 0$.

Case 5: For $W_{20} = 5$

For defocus aberration term $W_{20} = 5$, the transverse intensity distributions at shifted image plane due to self-similar members of Group I azimuthal Walsh filters for orders $\nu = 1, 3$ and 7 are illustrated in Fig. 5.6a–c respectively, for azimuth $\zeta = 0$.

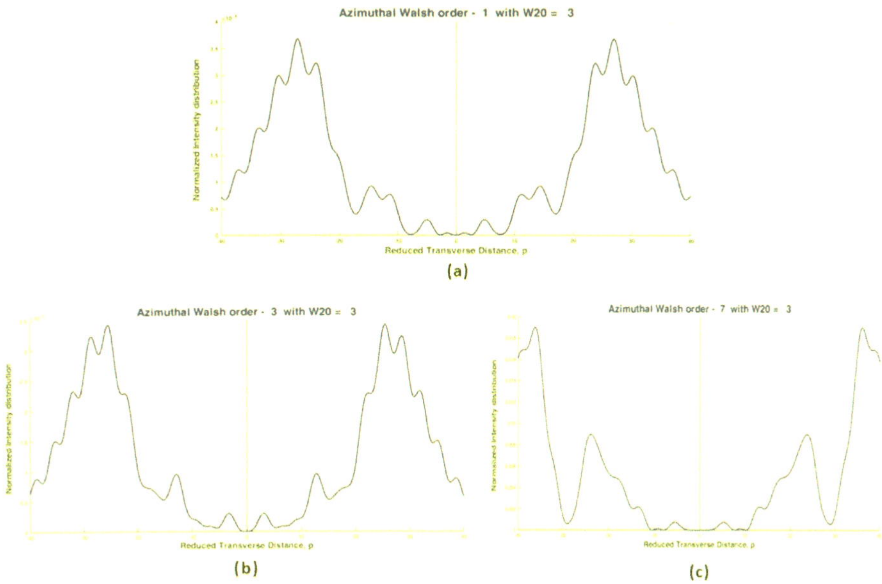

Fig. 5.4 **a-c** Transverse intensity distributions at shifted image plane with $W_{20} = 3$, due to self-similar members of Group I azimuthal Walsh filters for orders, $\nu = 1, 3$ and 7 respectively, for any particular azimuth $\zeta = 0$

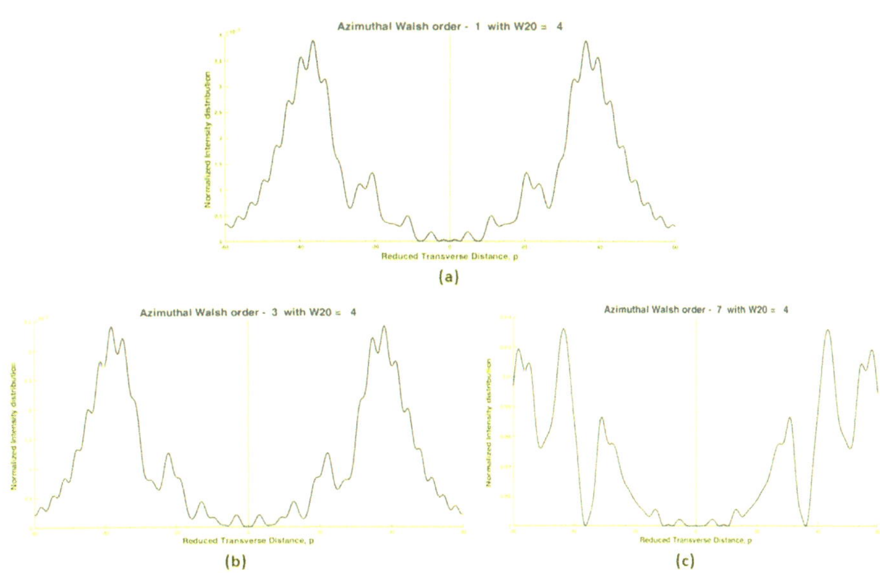

Fig. 5.5 **a–c** Transverse intensity distributions at shifted image plane with $W_{20} = 4$, due to self-similar members of Group I azimuthal Walsh filters for orders, $\nu = 1, 3$ and 7 respectively, for any particular azimuth $\zeta = 0$

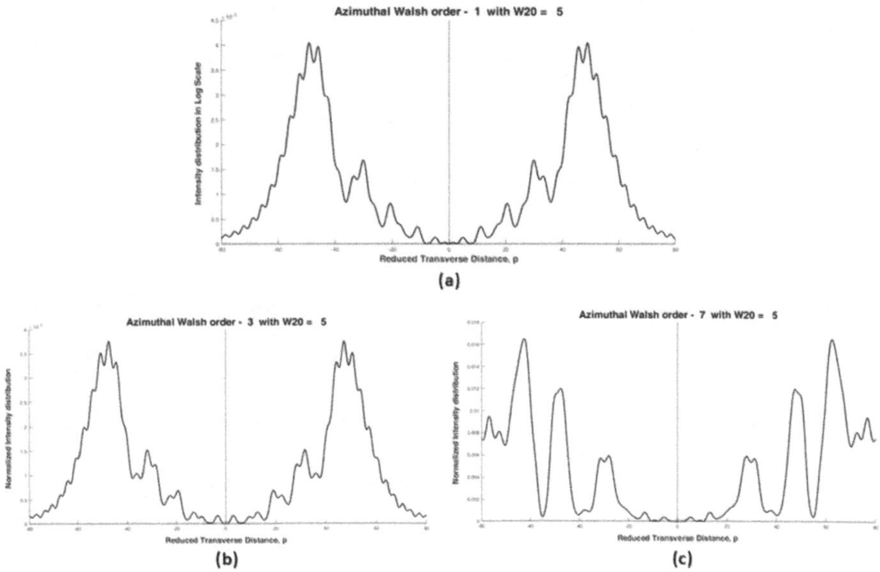

Fig. 5.6 **a–c** Transverse intensity distributions at shifted image plane with $W_{20} = 5$, due to self-similar members of Group I azimuthal Walsh filters for orders, $\nu = 1, 3$ and 7 respectively, for any particular azimuth $\zeta = 0$

5.2.2 Study of Self-Similarity on Transverse Image Planes Shifted Towards Left or (-)ve Side of Far-Field Plane

Study of self-similarity in the transverse intensity distributions on the longitudinally shifted image planes from the far-field plane by an amount—ΔZ, where—ΔZ represents longitudinal shift of the image planes from the far-field plane along (−)ve direction or towards the left, has been presented in this section. The corresponding defocus or 'Depth of Focus' aberration term W_{20} in this region has been considered to be within the range $-5 \leq W_{20} \leq 0$ with successive steps of -1.

Case 1: For $W_{20} = -1$
For a defocus aberration term $W_{20} = -1$, the transverse intensity distributions due to Group I members of azimuthal Walsh filters for orders $\nu = 1, 3$ and 7 and for azimuth $\zeta = 0$ are illustrated in Fig. 5.7a–c respectively.

Case 2: For $W_{20} = -2$
For a defocus aberration term $W_{20} = -2$, the transverse intensity distributions due to Group I members of azimuthal Walsh filters for orders $\nu = 1, 3$ and 7 and for azimuth $\zeta = 0$ are illustrated in Fig. 5.8a–c respectively.

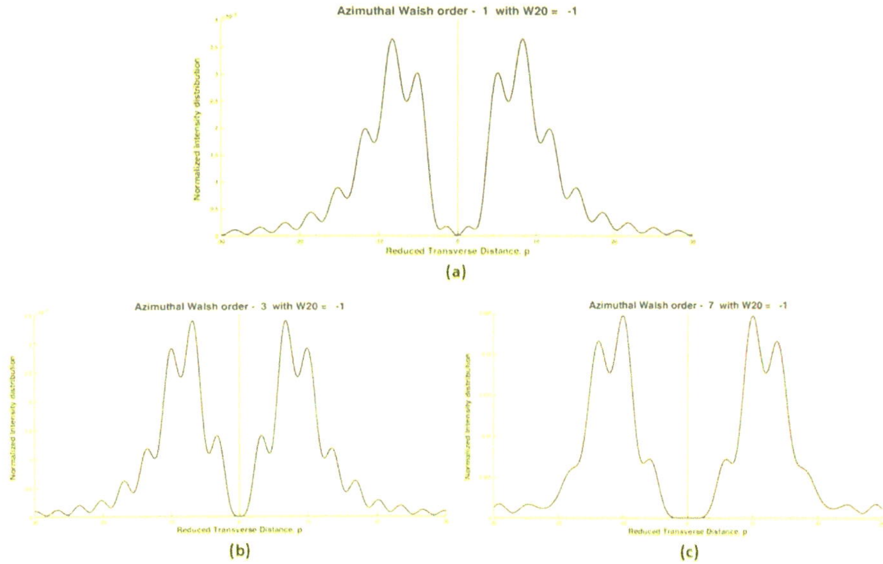

Fig. 5.7 **a–c** Transverse intensity distributions at shifted image plane with $W_{20} = -1$, due to self-similar members of Group I azimuthal Walsh filters for orders, $\nu = 1, 3$ and 7 respectively, for azimuth, $\zeta = 0$

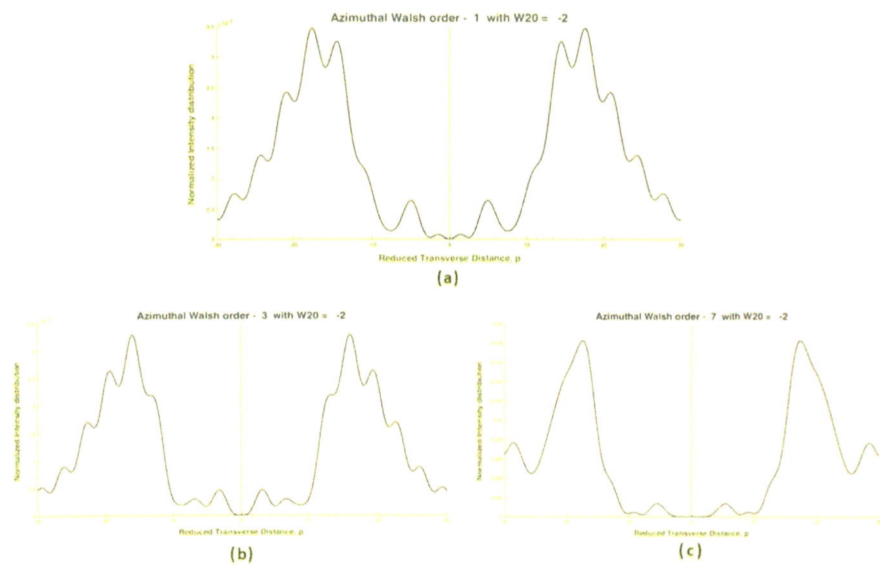

Fig. 5.8 **a–c** Transverse intensity distributions at shifted image plane with $W_{20} = -2$, due to self-similar members of Group I azimuthal Walsh filters for orders, $\nu = 1, 3$ and 7 respectively, for azimuth, $\zeta = 0$

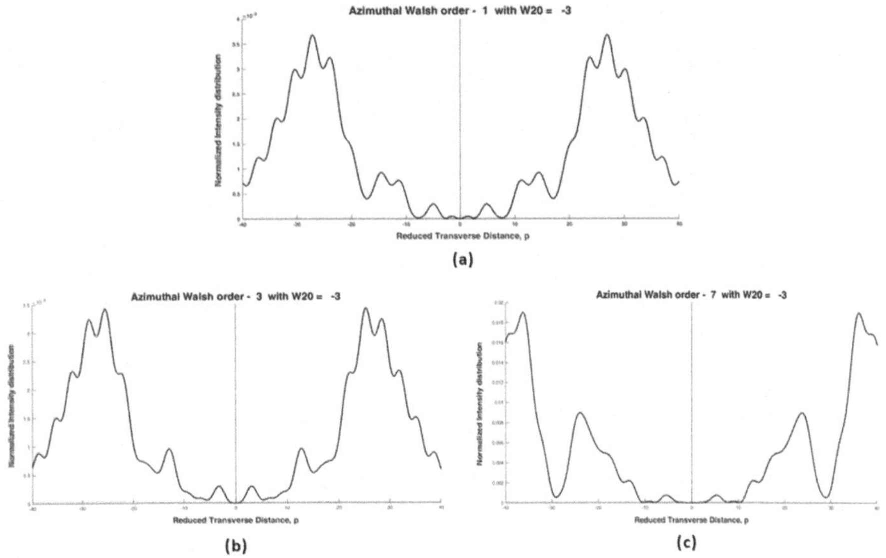

Fig. 5.9 a–c Transverse intensity distributions at shifted image plane with $W_{20} = -3$, due to self-similar members of Group I azimuthal Walsh filters for orders, $\nu = 1, 3$ and 7 respectively, for azimuth, $\zeta = 0$

Case 3: For $W_{20} = -3$
For a defocus aberration term $W_{20} = -3$, the transverse intensity distributions due to Group I members of azimuthal Walsh filters for orders $\nu = 1, 3$ and 7 and for azimuth $\zeta = 0$ are illustrated in Fig. 5.9a–c respectively.

Case 4: For $W_{20} = -4$
For a defocus aberration term $W_{20} = -4$, the transverse intensity distributions due to Group I members of azimuthal Walsh filters for orders $\nu = 1, 3$ and 7 and for azimuth $\zeta = 0$ are illustrated in Fig. 5.10a–c respectively.

Case 5: For $W_{20} = -5$
For a defocus aberration term $W_{20} = -5$, the transverse intensity distributions due to Group I members of azimuthal Walsh filters for orders $\nu = 1, 3$ and 7 and for azimuth $\zeta = 0$ are illustrated in Fig. 5.11a–c respectively.

It is observed from Figs. 5.2, 5.3, 5.4, 5.5, 5.6, 5.7, 5.8, 5.9, 5.10 and 5.11 that self-similar members of Group I azimuthal Walsh filters exhibit self-similarity in the transverse intensity distributions while computed on shifted image planes around far-field region. It is further observed that as the order ν increases, the self-similar intensity distributions show wider central minima. As the shift in image plane ΔZ increases with an increase in the corresponding wavefront aberration term W_{20}, for any particular order ν, the ripples or noise increases resulting in defocusing in the corresponding intensity distribution function. It is further observed that the self-similar

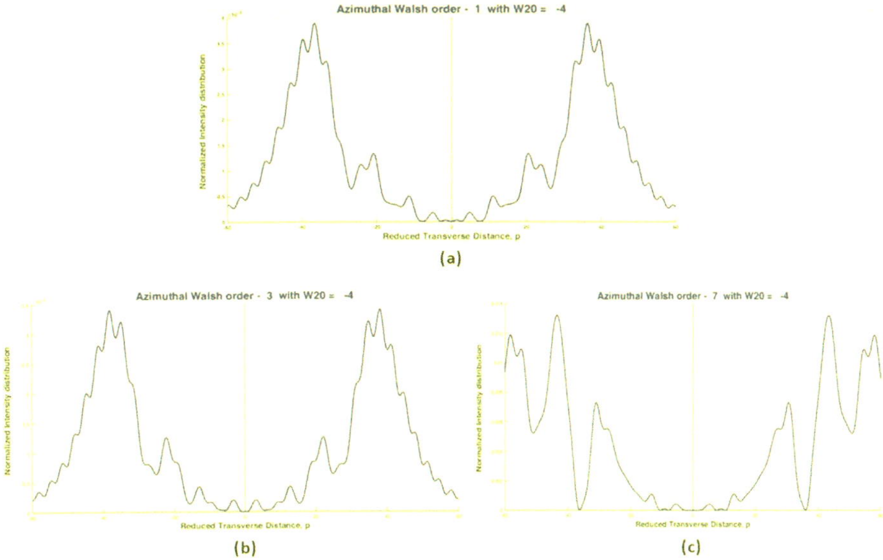

Fig. 5.10 **a–c** Transverse intensity distributions at shifted image plane with $W_{20} = -4$, due to self-similar members of Group I azimuthal Walsh filters for orders, $\nu = 1, 3$ and 7 respectively, for azimuth, $\zeta = 0$

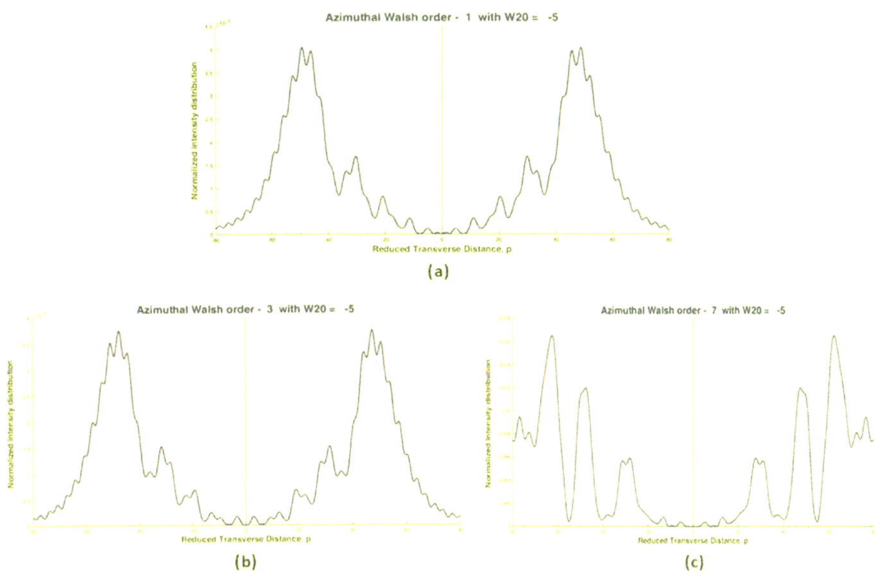

Fig. 5.11 **a–c** Transverse intensity distributions at shifted image plane with $W_{20} = -5$, due to self-similar members of Group I azimuthal Walsh filters for orders, $\nu = 1, 3$ and 7 respectively, for azimuth, $\zeta = 0$

intensity distributions are exactly identical for image planes shifted longitudinally by an amount $+\Delta Z$ or $-\Delta Z$, for any particular order ν.

5.3 Self-Similarity in Transverse Intensity Distributions in the Far-Field Region for Group IIA Self-Similar Members of Azimuthal Walsh Filters

Azimuthal Walsh filters of Group IIA consisting of orders, $\nu = 2, 4, \ldots$ are observed to produce self-similar transverse intensity distributions around the far-field region, when computed for any particular azimuth. Transverse intensity distributions around the far-field plane due to Group IIA self-similar members, for a particular azimuth, $\zeta = 0$, taking into consideration of defocus aberration term W_{20} within the ranges of $0 \leq W_{20} \leq +5$ and $-5 \leq W_{20} \leq 0$ have been illustrated in this section. The axial intensity on the shifted image planes is observed to be zero as observed in case of Group I self-similar members.

In absence of defocus aberration term, $W_{20} = 0$, at the far-field plane, the transverse intensity distributions, due to self-similar members of Group II A azimuthal Walsh filters, for orders $\nu = 2, 4, \ldots$, calculated for azimuth $\zeta = 0$, are illustrated in Fig. 5.12a, b respectively. It is seen from Fig. 5.12 that for $W_{20} = 0$ i.e., on the far-field plane, the normalized intensity distributions exhibited by Group IIB self-similar members of azimuthal Walsh filters show self-similar pattern with sudden central dip or minima. The dip widens with the increase of order number, ν.

5.3.1 Study of Self-Similarity on Transverse Image Planes Shifted Towards Right or (+)ve Side of Far-Field Plane

Study of self-similarity in transverse intensity distributions on the longitudinally shifted image planes from the far-field plane, by an amount $+\Delta Z$, where $+\Delta Z$ represents longitudinal shift of the image plane from the far-field plane along (+)ve direction, has been presented in this section. The corresponding defocus or 'Depth of Focus' aberration term W_{20} in this region has been within the range $0 \leq W_{20} \leq +5$ with successive steps of 1.

Case 1: For $W_{20} = 1$

For the value of defocus aberration term $W_{20} = 1$, the transverse intensity distributions due to Group IIA self-similar members of azimuthal Walsh filters for orders $\nu = 2$ and 4 are illustrated in Fig. 5.13a, b respectively, for azimuth $\zeta = 0$.

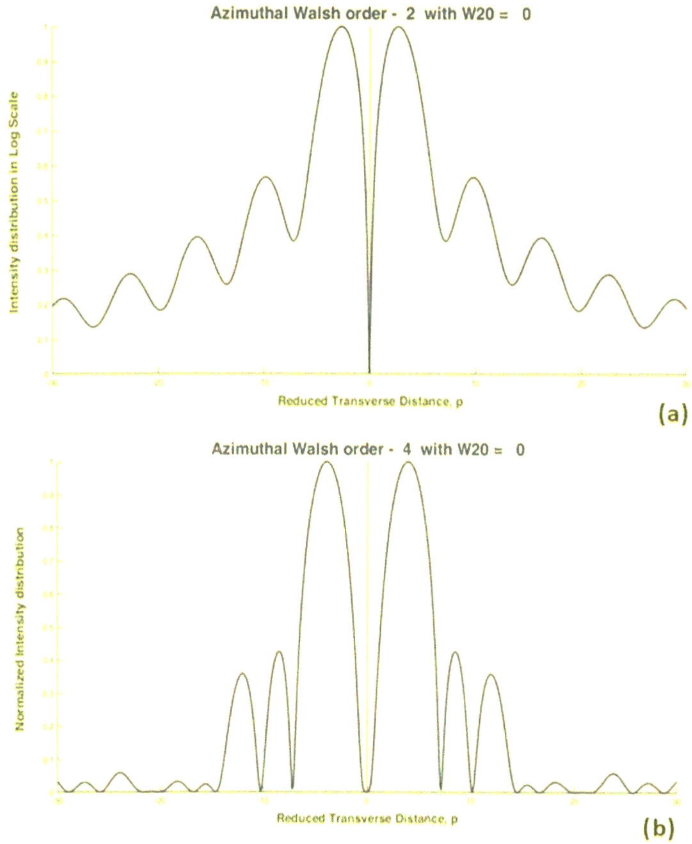

Fig. 5.12 **a**, **b** Transverse intensity distributions with $W_{20} = 0$, due to self-similar members of Group IIA for orders, $\nu = 2, 4, \ldots$ respectively, for azimuth $\zeta = 0$

Case 2: For $W_{20} = 2$

For the value of defocus aberration term $W_{20} = 2$, the transverse intensity distributions due to Group IIA self-similar members of azimuthal Walsh filters for orders $\nu = 2$ and 4 are illustrated in Fig. 5.14a, b respectively, for azimuth $\zeta = 0$.

Case 3: For $W_{20} = 3$

For the value of defocus aberration term $W_{20} = 3$, the transverse intensity distributions due to Group IIA self-similar members of azimuthal Walsh filters for orders $\nu = 2$ and 4 are illustrated in Fig. 5.15a, b respectively, for azimuth $\zeta = 0$.

Case 4: For $W_{20} = 4$

For the value of defocus aberration term $W_{20} = 4$, the transverse intensity distributions due to Group IIA self-similar members of azimuthal Walsh filters for orders $\nu = 2$ and 4 are illustrated in Fig. 5.16a, b respectively, for azimuth $\zeta = 0$.

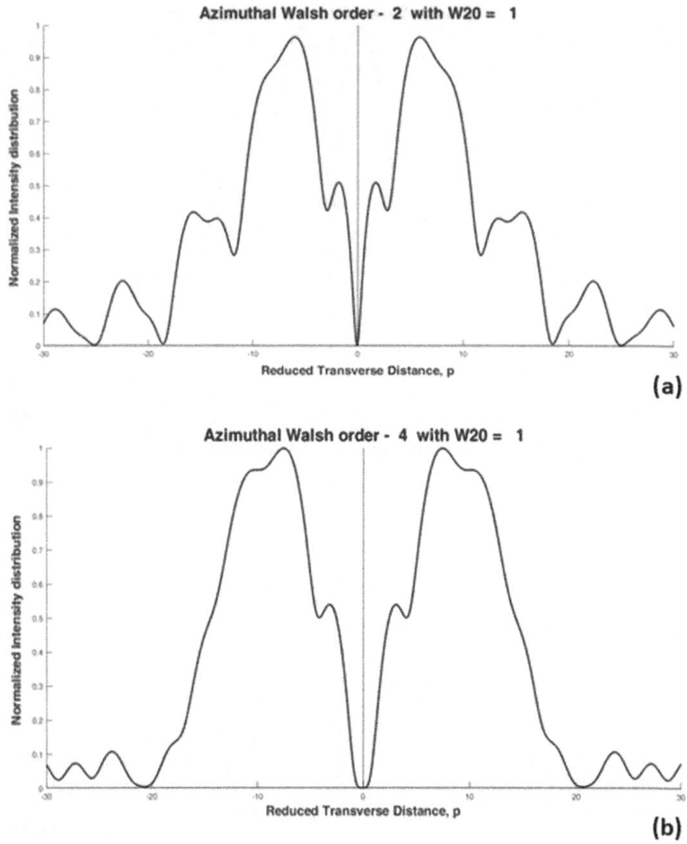

Fig. 5.13 **a**, **b** Transverse intensity distributions at shifted image planes around far-field region, with $W_{20} = 1$, due to self-similar members of Group IIA for orders, $\nu = 2, 4, \ldots$ respectively, for azimuth $\zeta = 0$

Case 5: For $W_{20} = 5$

For the value of defocus aberration term $W_{20} = 5$, the transverse intensity distributions due to Group IIA self-similar members of azimuthal Walsh filters for orders $\nu = 2$ and 4 are illustrated in Fig. 5.17a, b respectively, for azimuth $\zeta = 0$.

5.3.2 *Study of Self-Similarity on Transverse Image Planes Shifted Towards Left or (-)ve Side of Far-Field Plane*

Study of self-similarity in the transverse intensity distributions on the longitudinally shifted image planes from the far-field plane, by an amount—ΔZ, where—ΔZ represents the longitudinal shift of the image planes from the far-field plane along (-)ve

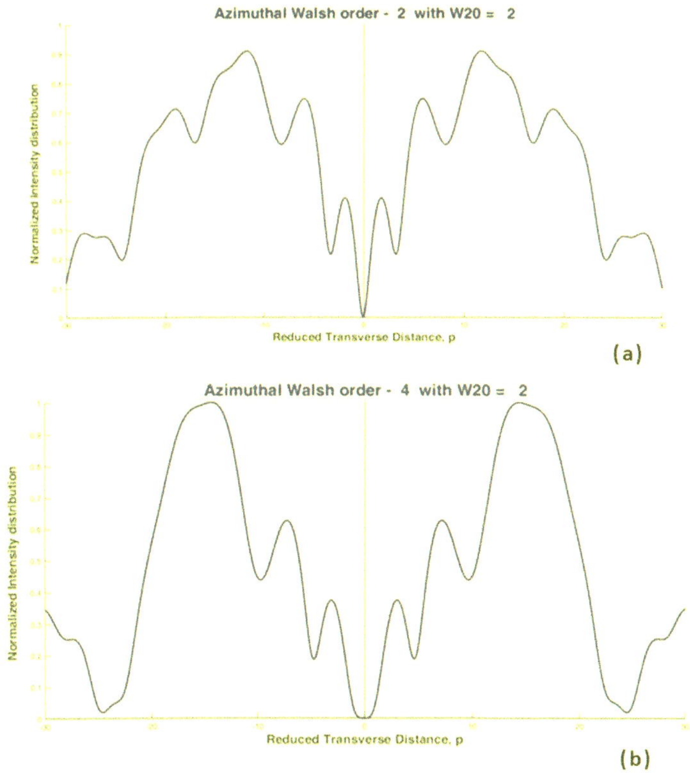

Fig. 5.14 **a**, **b** Transverse intensity distributions at shifted image planes around far-field region, with $W_{20} = 2$, due to self-similar members of Group IIA for orders, $\nu = 2, 4, \ldots$ respectively, for azimuth $\zeta = 0$

direction or towards the left, has been presented in this section. The corresponding defocus or 'Depth of Focus' aberration term W_{20} has been considered to be within the range of $-5 \leq W_{20} \leq 0$ with successive steps of -1.

Case 1: For $W_{20} = -1$

For a defocus aberration term $W_{20} = -1$, the transverse intensity distributions due to Group IIA members of azimuthal Walsh filters for orders $\nu = 2$ and 4 and for azimuth $\zeta = 0$, are illustrated in Fig. 5.18a, b respectively.

Case 2: For $W_{20} = -2$

For a defocus aberration term $W_{20} = -2$, the transverse intensity distributions due to Group IIA members of azimuthal Walsh filters for orders $\nu = 2$ and 4 and for azimuth $\zeta = 0$, are illustrated in Fig. 5.19a, b respectively.

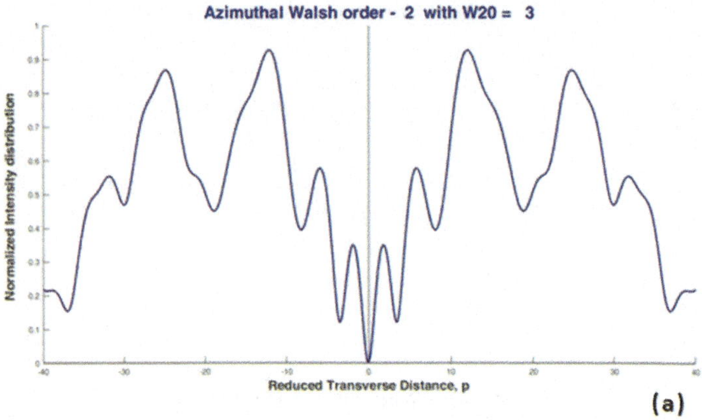

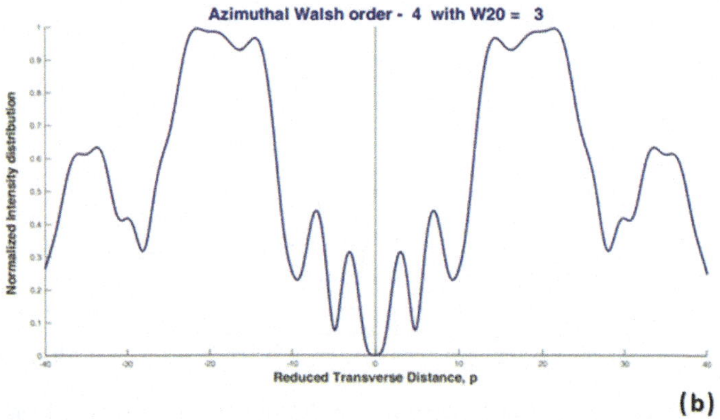

Fig. 5.15 **a, b** Transverse intensity distributions at shifted image planes around far-field region, with $W_{20} = 3$, due to self-similar members of Group IIA for orders, $\nu = 2, 4, \ldots$ respectively, for azimuth $\zeta = 0$

Case 3: For $W_{20} = -3$

For a defocus aberration term $W_{20} = -3$, the transverse intensity distributions due to Group IIA members of azimuthal Walsh filters for orders $\nu = 2$ and 4 and for azimuth $\zeta = 0$, are illustrated in Fig. 5.20a, b respectively.

Case 4: For $W_{20} = -4$

For a defocus aberration term $W_{20} = -4$, the transverse intensity distributions due to Group IIA members of azimuthal Walsh filters for orders $\nu = 2$ and 4 and for azimuth $\zeta = 0$, are illustrated in Fig. 5.21a, b respectively.

Case 5: For $W_{20} = -5$

For a defocus aberration term $W_{20} = -5$, the transverse intensity distributions due to Group IIA members of azimuthal Walsh filters for orders $\nu = 2$ and 4 and for azimuth $\zeta = 0$, are illustrated in Fig. 5.22a, b respectively.

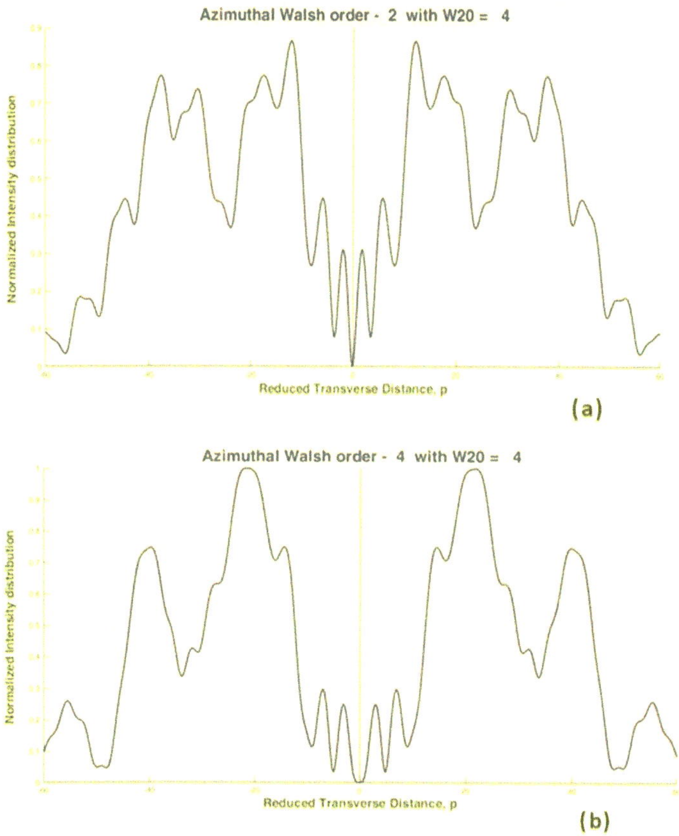

Fig. 5.16 **a**, **b** Transverse intensity distributions at shifted image planes around far-field region, with $W_{20} = 4$, due to self-similar members of Group IIA for orders, $\nu = 2, 4, \ldots$ respectively, for azimuth $\zeta = 0$

It is observed from Figs. 5.13, 5.14, 5.15, 5.16, 5.17, 5.18, 5.19, 5.20, 5.21 and 5.22 that self-similar members of Group II B azimuthal Walsh filters exhibit self-similarity in the transverse intensity distributions computed on shifted image planes around far-field region. It is further observed that as the order ν increases, the self-similar intensity distributions show wider central minima. As the shift in image plane ΔZ increases with an increase in the corresponding wavefront aberration term W_{20} for any particular order ν, the ripples or noise increases resulting in defocusing in the corresponding intensity distribution function. It is further observed that the self-similar intensity distributions are exactly identical for image planes shifted longitudinally by an amount $+\Delta Z$ or $-\Delta Z$, for any particular order ν.

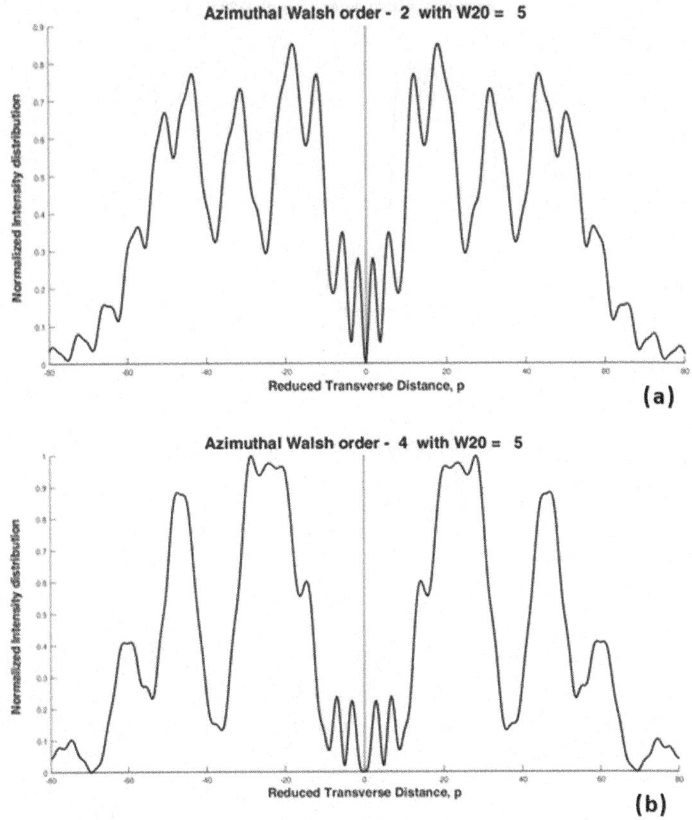

Fig. 5.17 **a**, **b** Transverse intensity distributions at shifted image planes around far-field region, with $W_{20} = 5$, due to self-similar members of Group IIA for orders, $\nu = 2$ and 4 respectively, for azimuth $\zeta = 0$

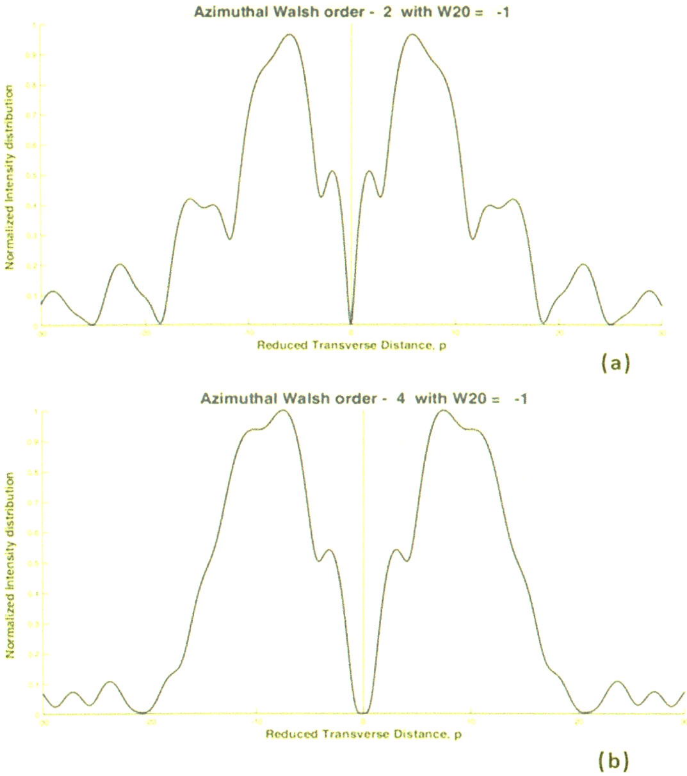

Fig. 5.18 **a**, **b** Transverse intensity distributions at shifted image planes around far-field region, with $W_{20} = -1$, due to self-similar members of Group IIA for orders, $\nu = 2$ and 4 respectively, for azimuth $\zeta = 0$

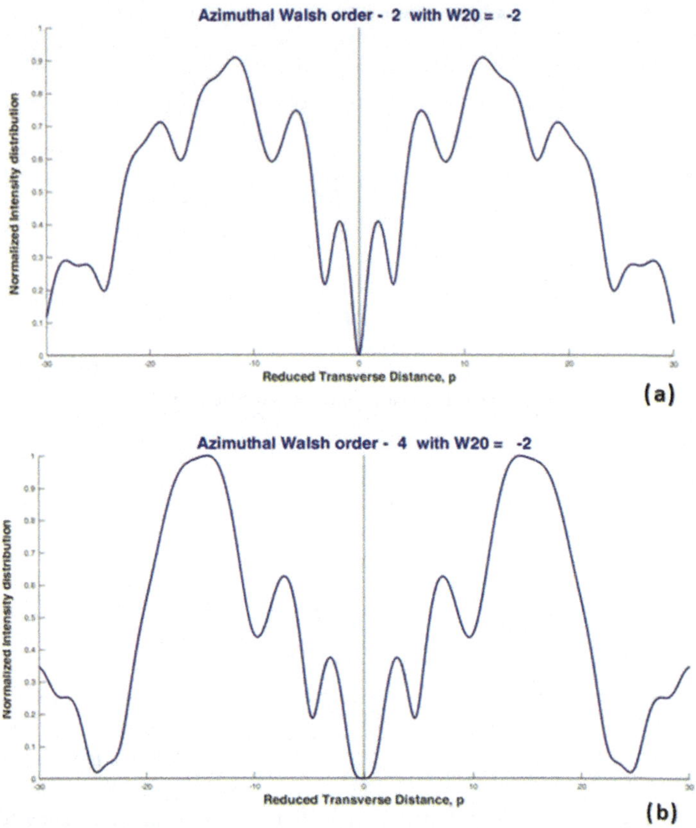

Fig. 5.19 **a**, **b** Transverse intensity distributions at shifted image planes around far-field region, with $W_{20} = -2$, due to self-similar members of Group IIA for orders, $\nu = 2$ and 4 respectively, for azimuth $\zeta = 0$

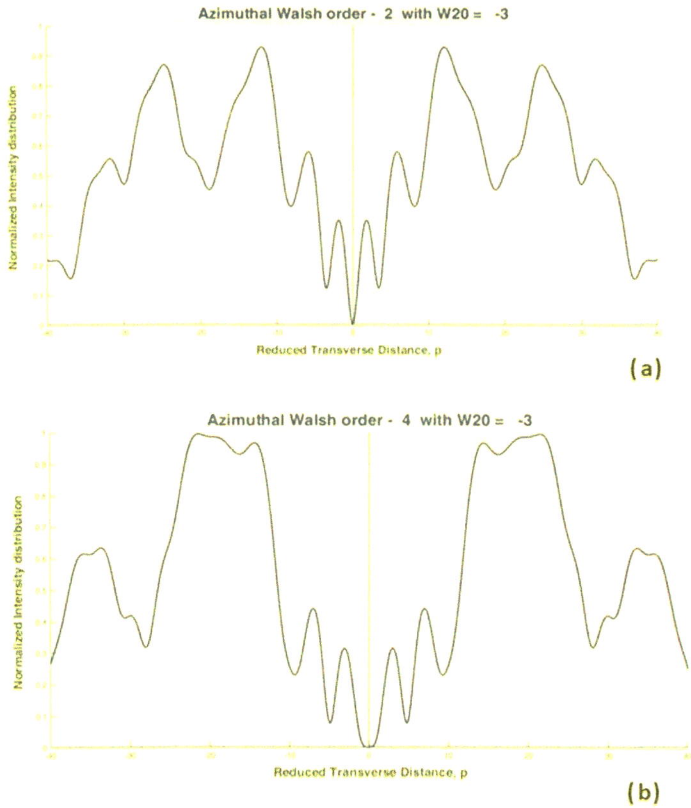

Fig. 5.20 **a**, **b** Transverse intensity distributions at shifted image planes around far-field region, with $W_{20} = -3$, due to self-similar members of Group IIA for orders, $\nu = 2$ and 4 respectively, for azimuth $\zeta = 0$

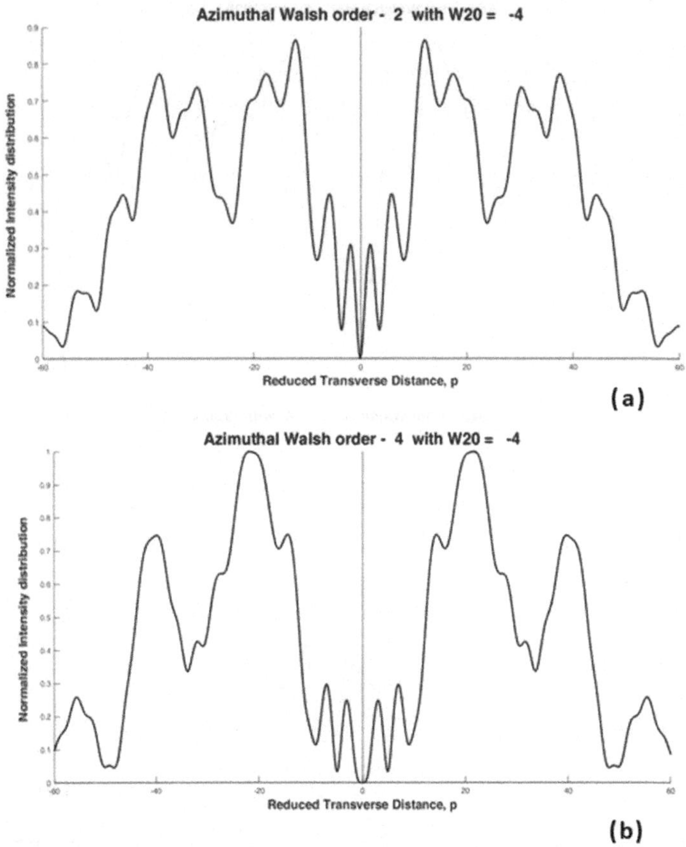

Fig. 5.21 **a**, **b** Transverse intensity distributions at shifted image planes around far-field region, with $W_{20} = -4$, due to self-similar members of Group IIA for orders, $\nu = 2$ and 4 respectively, for azimuth $\zeta = 0$

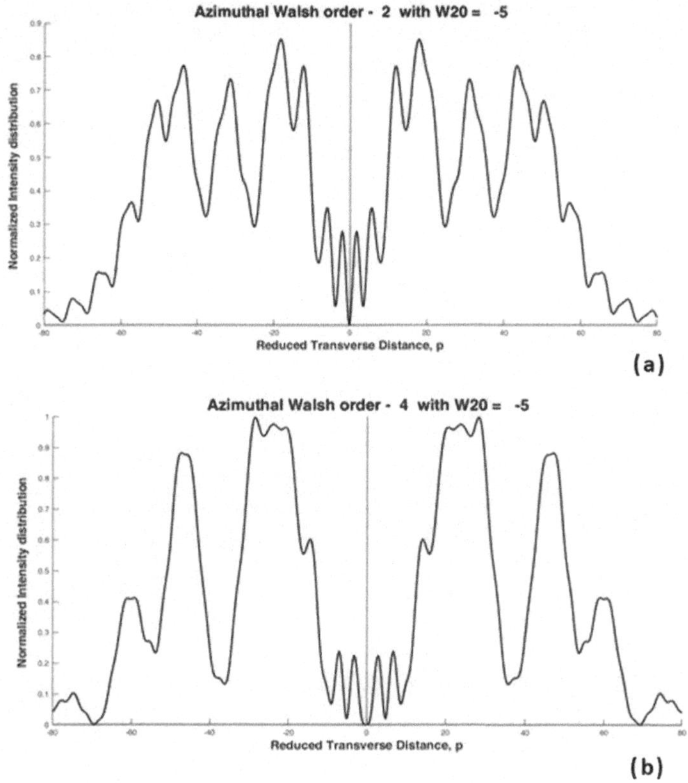

Fig. 5.22 **a, b** Transverse intensity distributions at shifted image planes around far-field region, with $W_{20} = -5$, due to self-similar members of Group IIA for orders, $\nu = 2$ and 4 respectively, for azimuth $\zeta = 0$

References

1. Bhattacharya I, Lakshminarayan H (2019) Azimuthal Walsh filters: an interesting tool to produce 2D and 3D light structures. In: Martínez-García A et al (eds.) Progress in optomechatronic technologies, vol 233. Springer Proceedings in Physics, pp 11–20
2. Bhattacharya I (2020), Self-similarity in azimuthal walsh filters and corresponding far-field diffraction characteristics: a unique study to control tightly focused fields and coupling of light into metamaterials, plasmonic structure and waveguides. In: proceedings of the SPIE 11257, plasmonics in biology and medicine XVII, vol 1125717, pp 1–14

Chapter 6
Future Perspectives

The possibility of using azimuthal Walsh filters, derived from the orthogonal set of azimuthal Walsh functions, as base functions have been proposed to manipulate the far-field diffraction patterns of imaging systems opens up new frontiers in wave optical engineering. The existence of self-similarity observed in the 2D intensity distributions exhibited by members of self-similar Groups and Sub-Groups of azimuthal Walsh filters unfolds another interesting feature of periodic or semi-periodic diffracting structures. The rotational self-similarity portrayed in the 2D intensity distributions by corresponding adjacent as well as distant orders of self-similar azimuthal Walsh filters rotated by specific angles provides a unique property exhibited by azimuthal Walsh filters compared to radial and annular categories. We have used MATLAB 2017a platform for the advance level computation and generate high-quality graphics for the study of the far-field diffraction pattern for different orders of azimuthal Walsh filters.

Further scope of research includes the use of azimuthal Walsh filters as phase-shifted zone plates to generate complex 3D beam structures like petal-shaped and optical ring lattice beams [1, 2]. Azimuthal Walsh filters can be used very efficiently near the focus of an imaging system to cater the needs of complex imaging in advanced microscopy, 3D imaging, lithography, optical super resolution, optical tomography and many other related fields. Optical micro- and nano-manipulation using azimuthal variant Walsh filters or combinations with radial variety can be used to form gradient force trap for optical tweezers [3, 4] for manipulating objects of size comparable to a single atom to 100 μm without mechanical contact.

It would be worthwhile to demonstrate the principal phenomenon of Optics to be presented by a collection of high-quality images and photographs, which is very much useful for the purpose of research and teaching. The effort has been explored by Cagnor et al. [5] for some specific optical phenomena supported by experiments. This research monograph may be referred to an atlas of diffraction phenomenon with

I. Bhattacharya and L. Hazra, *Azimuthal Walsh Filters*, Progress
in Optical Science and Photonics 10,
https://doi.org/10.1007/978-981-15-6099-6_6

sector-shaped apertures as pupil functions represented by azimuthal Walsh filters of different orders. Further scope of the research monograph lies in extending the investigation in case of two dimensional binary polar Walsh filters which may be formed by the combinations of Azimuthal and radial or annular varieties to cater into solutions for more general problems like generating multidimensional optical trap and tackling inverse problem where a phase filter needs to be synthesized in accordance with pre-specified diffraction characteristics.

References

1. S. Mukhopadhyay, S. Sarkar, K. Bhattacharya, L.N. Hazra, Polarisation phase shifting interferometric technique for phase calibration of a reflective phase spatial light modulator. Opt. Eng. **52**(3), 035602-1–035602-6 (2013)
2. A. Sabatyan, J. Rafighdoost, Azimuthal phase-shifted zone plates to produce petal-like beams and ring lattice structures. J. Opt. Soc. Am. A **34**(5), 919–922 (2017)
3. P.N. Lebedev, Experimental examination of light pressure. Ann. Phys. **6**, 433–458 (1901)
4. A. Ashkin, Acceleration and trapping of particles by radiation pressure. Phy. Rev. Lett. **24**(4), 156–159 (1970)
5. M. Cagnet, M. Françon, J.C. Thrierr, *Atlas of optical phenomena* (Springer, 1962), p. 1

The manufacturer's authorised representative in the EU is Springer Nature Customer Service Centre GmbH, Europaplatz 3, 69115 Heidelberg, Germany. If you have any concerns regarding our products, please contact ProductSafety@springernature.com

Printed and bound by CPI Group (UK) Ltd, Croydon, CR0 4YY
30/09/2026
02225515-0002